Chaos Theology

A Revised Creation Theology

Chaos Theology

A Revised Creation Theology

Sjoerd L. Bonting

© 2002 Novalis, Saint Paul University, Ottawa, Canada

Cover: Miriam Bloom
Layout: Caroline Gagnon

Business Office:
Novalis
49 Front Street East, 2nd Floor
Toronto, Ontario, Canada
M5E 1B3

Phone: 1-800-387-7164 or (416) 363-3303
Fax: 1-800-204-4140 or (416) 363-9409
E-mail: cservice@novalis.ca

National Library of Canada Cataloguing in Publication Data

Bonting, Sjoerd Lieuwe
    Chaos theology : a revised creation theology

(Saint Paul University research series)
Includes bibliographical references and index.
ISBN 2-89507-171-3

    1. Chaos (Christian theology) 2. Theodicy. 3. Evolution—Religious aspects—Christianity. I. Title. II. Series.

BL227.B66 2002   231.7'65   C2001-904103-9

Printed in Canada.

All rights reserved. No part of this publication may be reproduced, stored in a retrieval system, or transmitted in any form, or by any means, electronic, mechanical, photocopying, recording, or otherwise, without the written permission of the publisher.

We acknowledge the financial support of the Government of Canada through the Book Publishing Industry Development Program (BPIDP) for our publishing activities.

Contents

Foreword 7

Introduction 11

1. Chaos Theology as a New Approach 13
    1.1 A Novel View of Creation 14
    1.2 Problems with *Creatio ex Nihilo* 15
    1.3 Principles of Chaos Theology 20
    1.4 Chaos Theology and the Scientific World View ... 26
    1.5 Chaos Theory and Chaos Events 32
    1.6 Summary and Conclusions 36

2. God's Action in the World 39
    2.1 Influence of the Scientific World View 40
    2.2 Contingency in Cosmic Evolution 41
    2.3 Contingency in Biological Evolution 43
    2.4 Transcendent and Immanent Action of God 48
    2.5 Jesus Christ and Incarnation 55
    2.6 Jesus Christ and Reconciliation 57
    2.7 Summary and Conclusions 61

3. The Problem of Evil 65
    3.1 Evil: Prominent Topic of Discussion 66
    3.2 Moral Evil and Human Ambivalence 67
    3.3 Original Sin and Predestination 69

3.4	Evil in *Creatio ex Nihilo* Context	72
3.5	Evil in Chaos Theology	76
3.6	A Theology of Disease	77
3.7	Future and Destiny	82
3.8	Summary and Conclusions	85

Notes 89

Index 99

Foreword

It's as if she had seen the Promised Land "from over Jordan"! Several years ago, a dear friend of Saint Paul University donated a substantial sum to be used to promote "the study and wider recognition of the relationship between faith (the reflection on and the 'practice' of faith) and science (the exact [or hard] sciences and human [or soft] sciences), especially in its cultural aspects.

After a lengthy debate, the Rector of Saint Paul University, Dr. Dale M. Schlitt, O.M.I., and the Vice Rector, Academic, Dr. Achiel Peelman, O.M.I., decided that the University would sponsor four public lectures in the field of science and religion in the year 2000 and later publish expanded versions of the lectures. They engaged me to write the letters, make the travel arrangements, and act as the University's interface with the authors.

In developing our concept, we wrote the following statement:

> As the new millennium begins, and on a more sober note than much of the hyperbole surrounding that passage, Saint Paul University is proposing a series of four public lectures in the field of science and religion. The theme of these lectures, "Creation and Creature," will allow scope for our four principal speakers to explore both the scientific and the theological aspects of the theme, together with the nexus between them.
>
> On the one hand, the question of creation itself may raise the relationship between chaos theory and an understanding of Genesis 1. On the question of creation's creatures, particularly humankind, there will be scope for

exploration of "creatureliness," the *imago dei*, the Pauline proclamations [neither Jew nor Greek; neither male nor female], and the responsibility of humanity, as created, for the other creations of its Creator.

Four topics emerged in consultation with the four invited speakers: the possibility of a Christian theology of science; the question of creation out of chaos *vs.* creation *ex nihilo*; the evolution of human understanding of God; and "knowing God and nature in a postmodern world." In that order, the speakers were Dr. Donald Lococo, C.S.B., Assistant Professor of Christianity and Culture at the University of Saint Michael's College, Toronto; Rev. Dr. Sjoerd Bonting, formerly Professor of Biochemistry at the University of Nijmegen, Holland, and a priest of the Anglican Communion; Ms. Carol Albright, Chicago Regional Director of the Center for Theology and the Natural Sciences and former Executive Editor of *Zygon: Journal of Religion and Science*; and Dr. Jitse van der Meer, Professor of Biology at Redeemer College, Ancaster, Ontario, and founding director of the Pascal Centre for Advanced Studies in Faith and Science at Redeemer College.

The lectures took place in Spring and Autumn 2000. Regrettably, the woman whose gift made the series possible – and whose wish to remain anonymous we continue to honour – died just two weeks before the first lecture. But she was able to see, as if "from over Jordan," the land promised in return for her generosity by way of my letter to her, which set out for her the details about the proposed lecture series and plans for publication. Although many at Saint Paul University who knew her and valued her friendship miss her, the memory of her longstanding friendship and her devotion to the University will continue. These publications are the first installment on that memorial.

Foreword

In the development and implementation of this project a number of people have participated by giving counsel and direction, by lending particular expertise at various stages. I would like to acknowledge my indebtedness, as series co-ordinator, to them all. The Rector, Dr. Dale M. Schlitt, O.M.I., and the then Vice Rector, Academic, Dr. Achiel Peelman, O.M.I., were the prime movers and directors of the project. Ms. Lucie Laplante, Executive Secretary to the Rector, provided administrative support and kept things on track. Professors Heather Eaton and James Pambrun of the Faculty of Theology, Saint Paul University, provided input and advice in the project's initial stages. Kevin Burns (Commissioning Editor), Anne Louise Mahoney (Managing Editor), Caroline Gagnon and Suzanne Latourelle (Designers, Novalis Graphic Design Studio), all of Novalis Publishing, advised and worked on the various stages of publicity for the lectures and on manuscript preparation and publication for the series. I would like to express my gratitude to all of them.

Ivan Timonin
Series Editor
January 2002

Introduction

In addition to expressing my appreciation for the invitation to present one of the Series 2000 Lectures organized by Saint Paul University and for the hospitality received during my week-long stay in Ottawa, I am most grateful for the opportunity to offer a much expanded and updated version of my message to a wider public.

I am privileged as an Anglican scientist-theologian to be allowed the opportunity to set forth a revised creation theology, which I call chaos theology, and to show its usefulness for the science-theology dialogue and a reconsideration of some important theological topics, even though this means abandoning the traditional *creatio ex nihilo* doctrine.

My premise is that we have two world views, the theological and the scientific world views, both God-given. These two world views, therefore, need not clash, but taken together should give us a deeper view of reality, provided we keep in mind the limitations of each. The scientific world view gives answers to the "How?" questions, while the theological world view responds to the "Why?" questions. The theologian may criticize the scientist when he falls into reductionism ("it is nothing but..."), while the scientist may take to task the theologian who makes statements that entail mechanisms that conflict with our scientific insight. Such is the case with *creatio ex nihilo,* as I try to demonstrate.

My attempt has been to write for the non-specialist, the theologian who is not a scientist, the scientist who is not a theologian, and for all those who wish to deepen and strengthen their belief while living in a world that has come to be dominated by science and its applications. So I avoid jargon from both disciplines as much as possible. On the other hand, in my text I flip back and forth between science and theology, which some theologians find a bit disconcerting. However, I hope it is clear in the text from which discipline any particular statement comes.

The monograph is divided into three long chapters, subdivided into several sections. The first part deals with origin and problems of *creatio ex nihilo* and sets forth chaos theology as an alternative, as well as presenting the physical theory of chaos events, which I combine with the former. The second part applies these insights to the question of God's action in the world. The last part presents some applications of chaos theology to the problem of evil in the world and to our thinking about the future.

I am always eager to receive comments and criticism, for which purpose I include my e-mail address.

Sjoerd L. Bonting
January 2002
<s.l.bonting@wxs.nl>

Chapter 1

Chaos Theology as a New Approach

1.1 A Novel View of Creation

Rejecting an 1800-year-old, universally held doctrine is not a small matter. Yet, in this monograph I want to challenge the doctrine of creation out of nothing (*creatio ex nihilo*), and propose a revised doctrine of creation. My writing and thinking about the reconciliation between the world views of science and theology[1] led me to reconsider the doctrine of *creatio ex nihilo*. I found several reasons for abandoning this doctrine, which I describe in section 1.2. After presenting a revised doctrine of creation, which I call *chaos theology*,[2] in section 1.3, I illustrate its usefulness for the dialogue between the two world views in section 1.4. Next I explain how combining chaos theology with *chaos theory*, the current physical theory of so-called *chaos events* (section 1.5), can lead to a comprehensive creation theology with important applications for other aspects of theology: God's action in the world (chapter 2) and the problem of evil (chapter 3).

Until the end of the second century, the biblical concept of creation from an initial chaos was basically retained by the Christian community. The older creation story in Genesis speaks about a lifeless desert (Genesis 2:5-6), the later story about a formless void, darkness, waters (Genesis 1:2). The Hebrew word for this, *tohu wabohu,* is also used in Isaiah 34:11 and in Jeremiah 4:23 for chaos, waste and void. The early Fathers Justin (*c.* 150) and Clement of Alexandria (*c.* 200) still accepted this view. Clement pointed to a passage in the Wisdom of Solomon: "For your all-powerful hand, which created the world out of formless matter..." (Wisdom 11:17) and applied the Platonic ideas of an ultimate divine reality, the One who in creation overflows into the surrounding void, and of eternal, pre-existing matter.[3] Creation from initial chaos is also the common view in non-biblical creation stories;[4] when in a single case the term "nothing" is used, it refers only to the initial absence of the structures and beings seen in the present world.

The idea of creation out of nothing was introduced by Theophilus of Antioch (*c.* 185) in his battle against the teaching of Marcion (*c.* 160) and the Gnostics who, noticing the evil in the world, proposed that the material universe was created from pre-existing evil matter by a demiurge, a lower imperfect god.[5] For Marcion this lower god was the God of the Old Testament, over against whom stood the higher God of the New Testament, who sent Jesus to liberate humanity from the evil world and from matter. Such a concept clearly violated Judeo-Christian monotheism. Against the Platonic and Gnostic ideas of pre-existent, eternal matter, Theophilus maintained that it "would be nothing great if God had made the cosmos out of pre-existent matter."[6] Irenaeus (*c.* 190) agreed with him; contrary to the Gnostic belief in a plurality of divine beings he upheld the belief in the one true God of the Old Testament and New Testament, and he rejected the idea of pre-existing and eternal matter by claiming that God took from himself the matter for the things he created. By "matter" Irenaeus meant God's will and power. Cosmological questions hardly worried him: "As the Bible gives no information, it is not permissible to speculate about it as the gnostics do."[7] Later Augustine (*c.* 400) accepted *creatio ex nihilo,* which was thereafter almost universally adopted by the Church. It was dogmatically formulated at the Fourth Lateran Council (1215), and reaffirmed by the first Vatican Council (1870). It was not a point of discussion during the Reformation, nor has it since been rejected. Why? Its attractiveness is probably that it is so "neat" (no mystery of an unexplained initial chaos) and that it seems to emphasize God's omnipotence. Nevertheless *creatio ex nihilo* raises several problems, which I shall discuss in the next section.

1.2 Problems with *Creatio ex Nihilo*

The concept of *creatio ex nihilo* presents five serious problems: conceptual, biblical, scientific, theological, and philosophical (the origin and existence of evil).

Conceptual problem

The conceptual problem is that none of us can picture absolute nothingness. This may explain why many theologians, like Augustine and Karl Barth, employ the term *nihil* in a rather loose fashion. They consider it as a *nihil ontologicum,* an existing nothing rather than a *nihil negativum,* absolute nothing. However, an existing *nihil* is not essentially different from an initial chaos. The same is true if one says with John Polkinghorne that *creatio ex nihilo* is merely a "metaphysical" statement.[8] Therefore I shall adhere to a strict interpretation of *nihil* as the complete absence of matter, energy, physical laws, structure, and order.

Biblical problem

The biblical problem is that *creatio ex nihilo* conflicts with both creation accounts in Genesis. Claus Westermann writes, in his authoritative commentary on Genesis,

> Such an abstract idea (i.e., *creatio ex nihilo*) is foreign to both the language and thought of P (the unknown author of Genesis 1); it is clear that there can be here no question of a *creatio ex nihilo;* our query about the origin of matter is not answered; the idea of an initial chaos goes back to mythical and premythical thinking."[9]

The Hebrew term *bará* differs essentially from the English "create"; it means separating, shaping into a form, defining, excluding and giving individuality, rather than calling forth from nothing.

Four texts are usually quoted in support of a *creatio ex nihilo*: Job 26:7 ("He stretches out the north over the void, and hangs the earth upon nothing"), Romans 4:17 ("God who... calls into existence the things that do not exist"), Hebrews 11:3 ("what is seen was made from things that are not visible"), and 2 Maccabees 7:28 ("God did not make them out of things that existed"). These texts fit equally well with creation from initial chaos, and thus can hardly be seen as clear evidence for creation *ex nihilo*. The *ex nihilo* concept is foreign to the Bible, which conclusion is also reached by David Fergusson.[10]

Scientific problem

Physical science, whether classical, quantum mechanical or relativistic, cannot explain the origin of the universe from a *nihil*, defined as the absence of matter, energy, physical laws, structure, and order. This is the conclusion drawn by Mark Worthing from an extensive study.[11] Currently, the nearly universally accepted theory for cosmic beginnings is the big-bang theory, which assumes an initial explosion at time zero. This theory cannot say anything about the conditions before and during the explosion, but addresses only the events after the initial fraction of a second (see section 1.4). However, an initial explosion requires some form of pre-existing energy, matter, gravity, unknown physical laws, or a combination of these, which is not a *nihil*. Some physicists describe the cosmic origin as a "quantum fluctuation in a vacuum," but this does not constitute an initial *nihil*. As John Polkinghorne says, "a quantum vacuum is a hive of activity, full of fluctuations, of random comings-to-be and fadings-away, certainly not something which without great abuse of language could be called 'nothing.'"[12] Arthur Peacocke appears to agree when he says, "It was not just 'nothing at all' even if it was 'no thing'"![13] The ironic fact is that if science could explain a beginning of the world from a *nihil*, then there would be no place left for a Creator. Mark Worthing ends his survey with this conclusion: Any theory explaining how something has come from nothing must assume some preexisting laws or energy or quantum activity in order to have a credible theory. Nothing comes out of nothing.[14]

Theological problem

Explaining a cosmic origin from a true *nihil* causes theologians as much of a problem as it does scientists. Many theologians give up on a true *nihil*. Karl Barth tries to reconcile the initial chaos of Genesis 1:2 with "nothing" by assuming a *nihil privativum*, which he calls "*das Nichtige*," a "nothing" of things already existing, but not real before they were created.[15] Emil Brunner goes even further and basically abandons *creatio ex nihilo* when stating "there never was a 'nothing' alongside of God," and the meaning of the biblical words "create" and "creation" is that "God alone creates the world with no other co-operating factor; this expresses something which is utterly beyond all human understanding. What we know as creation is never *creatio ex nihilo*, it is always the shaping of some given material."[16]

However, as I said before, an existing *nihil* is not essentially different from an initial chaos. Paul Tillich realizes this when he states, "The *nihil* out of which God creates is...the undialectical negation of being."[17] Mark Worthing states that creation out of absolute nothingness is an impossibility.[18] He also rejects a creation out of God's own "substance" as leading to a pantheistic deification of the physical world, but seems to come close to this in his final conclusion: "*Creatio ex nihilo*, therefore, signifies the theological recognition that God created a universe distinct from the divine being, not out of any preexisting matter or principle, but out of nothing other than the fullness of God's own being."[19] Oxford theologian Keith Ward rightly distinguishes between "origin" in the cosmological sense and "creation" in the theological sense and argues the case for a created universe, but does not discuss, much less explain, *creatio ex nihilo*, in the section in his book headed "Creation out of Nothing."[20]

Jürgen Moltmann has made a serious attempt to provide a theological explanation for a true *creatio ex nihilo*.[21] The first problem to be solved, he says, is where to locate an initial "nothingness." Initially, "it" must be inside God, so as not to limit his omnipresence; but for creation "it" must be externalized to avoid pantheistic deification of the created world. He then invokes two ideas from the Jewish kabbala (mysticism): *zimsum* and *shekinah*. The former is God's concentration and contraction, a withdrawing of God into himself; the latter says that the infinite God can so contract his presence that he can dwell in the temple. Moltmann combines these concepts, stating, "Where God withdraws himself from himself to himself, he can call something forth which is not divine essence or divine being." This something he calls a *nihil*, a God-forsaken space, hell, absolute death, annihilating nothingness, in which God establishes his creation. Next Moltmann applies Paul's idea of *kenosis*, God's self-emptying in the incarnation of Christ (Philippians 2:5-8), and the idea of God's self-humiliation in Christ's death on the cross. This leads him to these statements: (a) God withdraws *into* himself in order to go *out of* himself in creation; (b) if God is creatively active in the "nothing" which he has ceded and conceded, then the resulting creation still remains in God who has yielded up the initial "nothing" in himself; (c) the initial self-limitation of God, which permits creation, then assumes the glorious, unrestricted boundlessness in which the whole creation is transfigured; and (d) in relating initial creation to eschatological creation, the death of Christ overcomes the "annihilating nothingness, which persists in sin and death."[22]

Against Moltmann's rather obscure reasoning, which Fergusson calls "ultimately unconvincing,"[23] I have the following objections: (a) The attraction of a strict *creatio ex nihilo*, as emphasizing the absolute creative power of the Creator, is largely negated by the need to assume in creation *zimsum, shekinah, kenosis,* and God's self-humiliation; (b) The latter two concepts have traditionally been used for the incarnation and crucifixion of Christ (Philippians 2:7, 8), not for creation; (c) An "annihilating nothingness, which persists in sin and death" is no less mysterious than an unexplained initial chaos; moreover, sin and death is not falling into *nihil*, but rather falling into chaos; (4) A nothingness that annihilates can hardly be considered a true *nihil*, and a relaxed *nihil* is not essentially different from an initial chaos, as explained before; and (e) Finally, creation from chaos is the biblical concept of creation.

The problem of evil

Creation out of nothing would imply that God is also responsible for the evil in this world; physical evil in natural disasters and disease, moral evil that we humans commit. This monstrous thought has never been resolved in 1800 years of *creatio ex nihilo*. In recent years the problem has been honestly faced by several theologians.[24] However, none can solve the problem, not with Irenaeus' doctrine of original sin, nor with the Augustinian idea of *privatio boni* (evil is the absence of the good), nor with the Irenaean idea of evil as falling within God's good purpose. Brian Hebblethwaite closes his perceptive book with the plaintive words: "But even if these things are so (happiness bestowed by God in the end), there still remains the problem of the cost of human suffering and wickedness here on earth. It is very hard to reach a balanced view on this problem."[25] The Gnostics with their evil demiurge and pre-existent evil matter offered at least an explanation. This topic will be discussed in more detail in Chapter 3.

In view of the problems described above, I prefer to retain the biblical concept of creation from chaos. In the next section I expand the idea of creation from chaos into a "chaos theology" that can facilitate the reconciliation of the theological and scientific accounts of the origin of the universe and our place in it. It can also clarify some important theological topics like God's action in the world, Christ's work of salvation (Chapter 2), and the problem of evil (Chapter 3).

1.3 Principles of Chaos Theology

Creation from chaos

The Genesis 1 story, written after the Babylonian exile, contains elements of the Babylonian creation story, *Enuma Elish*, in particular the primordial watery chaos.[26] The wind (storm) or spirit of God (*ruach*, Genesis 1:2) superficially resembles the winds by which Marduk overcame Tiamat in *Enuma Elish*. There are, however, crucial and radical differences. While *Enuma Elish* places the chaos at the beginning, Genesis 1 speaks about God first. The Creator does not issue from the chaos; he is not limited by it, as are the gods in *Enuma Elish*. Also in contrast to *Enuma Elish*, Genesis 1 does not have a theogony, a creation of gods. The Creator in Genesis is outside and above matter and the process of creation. He is absolute and timeless; in the words of Isaiah, "I am the first and I am the last; besides me there is no god" (Isaiah 44:6). Another striking difference is that in Genesis 1 the one God turns initial chaos into created order by his sovereign Word. Yet, the chaos was there when God began to create: "In the beginning when God created the heavens and the earth, the earth was a formless void and darkness covered the face of the deep" (Genesis 1:1-2).[27] The author of Genesis 1 leaves this unexplained.

An interesting point is that virtually all non-biblical creation stories describe creation from initial chaos.[28] This concept apparently has the character of an *archetype* in the Jungian sense, a primordial image in the collective human unconscious. Furthermore, a prominent aspect in primitive religion is the distinction between the sacred and the profane.[29] The sacred is the world of reality, so a village is laid out in a manner that imitates a divine model and thus it participates in sacred reality. The space outside the village, the jungle, is considered profane, because it is not ordered according to the divine model; it is, in my terms, "remaining chaos." The sacred can serve as a principle of order, because it possesses the power to order. The continued ordering of the sacred space of the village is necessary in order to prevent it from being overwhelmed by the disorder of the profane, the surrounding jungle. This is the contingency of creation.

Remaining chaos and continuing creation

God creates not by destroying chaos, but by ordering it, by pushing back chaos in three separations (Genesis 1:2-10). God assigns boundaries to the primeval sea (Job 38:8-11; Psalm 104:7-9; Proverbs 8:27-31; Jeremiah 5:22), he sets a guard over the sea (Job 7:12), he orders the waters back (Psalm 18:15; Psalm 89:9), he stills the raging of the sea and the madness of the peoples (Psalm 65:7), and "he rebukes the sea and makes it dry" (Nahum 1:4). This implies that an element of chaos, often symbolized as "sea," remains in the created universe. This may also be reflected in Genesis 1:6 ("let it separate the waters from the waters"). In the vision of John, this remaining element of chaos will be abolished on the last Day: "I saw a new heaven and a new earth... *and the sea was no more"* (Revelation 21:1).

Each act of creation (except that of the firmament and the separation of the waters) is followed by the statement "And God saw that it was good." Often this phrase is taken to mean that the initial creation was perfect and complete. However, *tov,* the Hebrew word for "good," means good for its purpose, for its function, rather than good in actuality. The presentation of creation in six days plus a day of rest, which is not found in any other ancient creation myth, suggests a continuation of the creation process *(creatio continua)* towards a transcendent goal, the destiny of creation.[30] This is even more evident when we consider the first eleven chapters of Genesis as the full creation story, which then becomes a pre-scientific account of cosmic, biological, and social evolution guided by a transcendent and immanent Creator. During this time the creation is not yet perfected and contains a remaining element of chaos. My thesis and the centre of the chaos theology is that *the remaining element of chaos in the creation expresses itself in the evil in the world, both physical (natural disasters and disease) and moral (human evil)*.

Chaos and contingency

Contingency means being dependent, endangered and accidental. The contingency of creation, its absolute dependence on God, is a generally held theological belief. For many theologians this is the primary ground for the *ex nihilo* postulate. Philip Hefner even insists that *creatio ex nihilo* has less to do with origins than with dependence.[31] In my opinion the remaining element of chaos offers a better explanation for the contingency of the creation. The element of chaos presents a lasting threat to the world, so that Paul can say that "the whole creation has been groaning in labour pains until now" and "waits with eager longing" for its final liberation (Romans 8:19-22). The cosmos is continually moving between the poles of chaos and order. The scientist perceives this contingency in the unpredictability and accidentalness of cosmic and biological evolution (sections 2.2 and 2.3). The theologian concludes that the created universe is contingent in that it is forever in need of the support of the Creator's will.[32]

Chaos and evil

Since contingency also includes the element of being threatened, of imperfection and evil, Augustine and Barth connect contingency with the evil in creation, the former speaking about *privatio boni* (the absence of goodness), the latter about *das Nichtige*.[33] But this leaves unanswered the question of how there

1: Chaos Theology as a New Approach

can be imperfection and evil in a creation out of nothing, in which *everything* is created by a good and perfect Creator. Theologians through the ages have vainly sought an explanation for *theodicy* – the vindication or defence of a good and almighty God in view of the existence of evil (section 3.4), but, in the *creatio ex nihilo*, context this problem must remain insolvable.

In my opinion, however, chaos theology offers a reasonable explanation, which can also be of great pastoral value (section 3.6). The remaining element of chaos is expressed in evil – the physical evil of natural disasters and illness – as well as the moral evil that humans commit against each other and against nature (ecological crisis). Evil is not created, but is inherent in the remaining element of chaos. It is a characteristic of the ongoing creation (*creatio continua*), in which remaining chaos is pushed back and ordered until on the last Day it will be definitively abolished.

How can evil come from remaining chaos? I consider chaos itself as morally neutral. However, both humans and nature are under its influence, and this may lead to moral and physical evil, e.g., "chaotic thinking" may lead humans to evil behaviour. Paul seems to express this in Romans 7:15 ("I do not understand my own actions. For I do not do what I want, but I do the very thing that I hate"). In section 3.6 I explain that diseases like cancer and mental illness can be seen as an expression of remaining chaos in the universe. The same can be said for natural disasters such as earthquakes, volcanic eruptions, cyclones and tornadoes. On the other hand, chaos also has the potential for good. God, in his freedom and creativity, creates by ordering chaos. Likewise, we humans can to some extent order chaos through the use of our God-given freedom and creativity, as in art, science and technology.

Chaos and incarnation

The doctrine of the incarnation affirms that, at a definite time in history and a definite place in the universe, the eternal and pre-existing Word of God, the creative Logos, was implanted in Jesus of Nazareth, who thus became Jesus Christ, at once fully God and fully human. Paul describes the incarnation in terms of *kenosis,* God's self-emptying (Philippians 2:6-8: "God emptied himself, taking the form of a slave"). The incarnation is a cosmic event, in that the incarnate One is the cosmic Christ, who is the universal Saviour for the entire cosmos, not just for humans (Ephesians 1:20-23; Colossians 1:15-20). It is the necessary introduction to the final fulfillment of the continuing creation process, in Jesus' words, the coming of the New Kingdom. I return to this in section 2.5.

Jesus Christ is not affected by the chaos element as we are. Although fully human, he is said to be without sin (2 Corinthians 5:21: "For our sake he made him to be sin who knew no sin"; Hebrews 4:15: "…one who in every respect has been tempted as we are, yet without sin"). In this way Jesus can be said to be the new Adam (Romans 5:14; 1 Corinthians 15:45), although I do not wish to accept the idea of an initial human immortality that some read into the Genesis story of the Fall (particularly, Genesis 3:19). Paul's words in 1 Corinthians 15:21 ("For since death came through a human being, the resurrection of the dead has also come through a human being") seem to me to point to the forfeiting of a future eternal life rather than to the loss of an initial immortality. The latter is, of course, in direct conflict with the scientific evidence for biological evolution, which could only take place through the operation of the life cycle in all living beings, including humans.

Chaos and eschatology

Eventually, at the end of time, God will perfect his creation as foreseen by the Old Testament prophets. The belief in a new creation, "the new heaven and the new earth," dominates the message of second Isaiah (Isaiah 41:17-20; 43:18-21; 66:22), and leads to the expectation of a new heart and a new covenant (Jeremiah 31:31-34; Ezekiel 36:26-28; Hosea 2:18-23). In this new creation, the Creator's original intention, threatened by the power of chaos and frustrated by the rebellion of his human creatures, will be fulfilled. This idea culminates in the New Testament in the vision of John: "Then I saw a new heaven and a new earth; for the first heaven and the first earth had passed away, and the sea was no more" (Revelation 21:1). We can see this as the definitive removal of the remaining chaos element, marking the completion

1: Chaos Theology as a New Approach

of the continuing creation. The phrase "and the sea was no more" (Revelation 21:1) implies that this element of chaos will be completely abolished on the last day at the triumphant return of Christ.[34] The abolishing of the chaos element reflects the perfection and fulfillment of the present world, rather than its cataclysmic destruction and replacement. I return to this in section 3.7.

Some critical questions considered

To summarize, chaos theology consists of three points:

(a) Initial creation from an unexplained initial chaos;

(b) Continuing creation in God's battle with the remaining element of chaos; and

(c) Evil originating from the remaining element of chaos.

Here I consider five critical questions that may be raised:

1. *Can one abandon the doctrine of* creatio ex nihilo, *which has been nearly universally held since the third century?* As an Anglican I hold to the Anglican tripos (or tripod) of Bible, Tradition (as expressed in the ancient Creeds) and Reason (with which to consider the first two). I have shown that *creatio ex nihilo* is not biblical; it is not contained in the ancient creeds. It is thus part of the ongoing tradition of the Church, which is not unchangeable.

2. *Does creation from initial chaos re-introduce Gnostic dualism?* I do not think so, because I do not invoke a demiurge but maintain with Genesis 1 the absolute sovereignty of God who creates by his authoritative Word. Neither is initial chaos equivalent with the eternal matter of the Gnostics; it is a state rather than matter. The dualism between order and chaos is, like that between good and evil, light and dark, belief and unbelief, particle and wave, simply the recognition of a characteristic of the universe in which we live.

3. *Does the idea of God battling remaining chaos diminish God's omnipotence?* Bearing in mind that omnipotence is a vague speculative concept, I feel that a God who is battling remaining chaos till the final victory on the last day is more powerful than a

Creator who allows his initial creation to be spoiled by wayward humans, as Origen and Augustine suggest.

4. *Who created initial chaos, if not God?* This is the type of question we cannot meaningfully ask, because here we encounter the initial mystery. In section 1.4 I shall come back to this.

5. *Am I embracing "process theology" in the chaos theology?* The fact that the process theologians also happen to reject *creatio ex nihilo*[35] does not make me one of them, because I firmly reject their notion of an evolving Creator.

1.4 Chaos Theology and the Scientific World View

In this section I present some illustrations of the way in which chaos theology can facilitate the dialogue between the two world views of science and theology. While the Bible is certainly not a science textbook it can be said, in the words of Old Testament scholar Th.C. Vriezen, that "Genesis 1 shows marks of profound reflection in the field of religion as well as that of natural science."[36] It is therefore legitimate and useful to consider, in the light of the Genesis 1 story, to what extent the description of the reality of our world by each world view can be reconciled and integrated. In this way we may hope to obtain a deeper understanding of this reality and acquire a faith to live by in our time.

Initial mystery

Both science and theology are confronted with initial mystery. In the big-bang theory the entire course of 15 billion years of cosmic evolution can be calculated backwards from the present state of the universe until a point 10^{-43} seconds after time zero. Although this so-called Planck time is a very small fraction of a second indeed (10^{-43} seconds stands for 1 divided by 10 followed by 42 zeros), it is a decisive gap nevertheless. Quantum theory does not allow us to get any closer to time zero. Extrapolation to time zero only leads to a "singular" point with an infinitely small, infinitely dense and infinitely hot "mysterium." At this singularity the big bang took place and time (our chronological time)

1: Chaos Theology as a New Approach

began, but the theory cannot say anything about conditions and origin of the initial state. This is the initial mystery of the big bang.

The creation stories in Genesis also leave us with an initial mystery, the mystery of the initial chaos: a lifeless desert (Genesis 2:5-6) or a watery void (Genesis 1:2). Again, we are not informed about the conditions and origin of this initial chaos. Both theologian and scientist must admit the limitations of our understanding of beginnings. *"In the beginning"* (Genesis 1:1) finds its parallel in the scientific idea that our time began at the moment of the big bang. Creation in six successive "days" followed by a "day" of rest suggests a continuation of the creation process. This finds a parallel in the scientific view of cosmic and biological evolution.

If we consider with most Old Testament scholars the first eleven chapters of Genesis to be the full creation story, then we obtain therein a pre-scientific account of cosmic, biological, and social evolution guided by a transcendent and immanent Creator. His transcendence is manifested in big bang and institution of the physical laws (initial creation), his immanence in the subsequent 15 billion years of cosmic and biological evolution (continuing creation). To those theologians who maintain that the Genesis stories are merely meant to proclaim that the universe was created by God, I say that then the Genesis authors could have limited themselves to such a statement. To me it seems obvious that they were not only interested in the "Why" question, but also in the "How" question, which nowadays is the province of the scientists. Thus it is right and desirable to consider our current scientific insight in the exegesis of the Genesis creation stories.

Separation and ordering

The Genesis 1 creation story presents a process of separation and ordering. First it speaks about three separations (Genesis 1:3-10): light and darkness; water and heaven; earth and sea. Then follows the ordering, when by his creative Word God turns primordial chaos into created order: plants, heavenly bodies, animals, and humans (Genesis 1:11-31). The sequence has a striking resemblance to that assumed in modern cosmology and biological evolution, except that the plants appear before the Sun: author P was apparently oblivious of the process of photosynthesis! The element of physical and moral evil in our world is introduced by P in Genesis 6:9-22 with the story of the Flood, and by author J of the older creation account in the story of the Fall in Genesis 3. As mentioned before, I attribute the presence of evil to the remaining chaos element in creation (chapter 3).

The big-bang theory also has three separations at the beginning: the separation of time and space; the separation of the four fundamental forces (gravity, strong and weak nuclear forces, electromagnetic force); and the separation of the elementary particles (in first instance electrons, quarks and gluons, the latter two turning quickly into protons and neutrons). These separations set the stage for a process of ordering, leading to the appearance of galaxies, stars and planets, and, on planet Earth, plants, animals, and humans.

Some other analogies can be mentioned.[37] The conclusion that time began with the big bang finds an analogy in the statement by Augustine that the universe was created "with" rather than "in" time. The cosmological insight that the universe has no centre has a counterpart in the theological insight that God is everywhere and is not limited to one location. The fact that the entire cosmos was required to enable the emergence of humans on planet Earth is reflected in the unique place assigned to humans in Genesis 1.

Notwithstanding these analogies, the Genesis 1 story is primarily meant as a reflection about the relation between God, world, and humankind. The creation story can give a meaning and purpose to the process of cosmic and biological evolution, which science by its nature cannot provide and which is neglected in the "nothing but" and "chance only" ideology of some scientists. The presence of evil in the created world, for which science cannot give a satisfactory explanation, can be explained as the result of the operation of the remaining element of chaos in creation.

1: Chaos Theology as a New Approach

Chaos and entropy

The second law of thermodynamics states that every closed system left to itself will in the course of time increase its *entropy*, the thermodynamic measure of disorder. The production of galaxies and stars during cosmic evolution, and of living organisms during biological evolution, does not violate this law.[38] These structures and organisms are open systems that exchange energy and matter with their surroundings. Every living organism on Earth receives energy directly or indirectly from the Sun, takes up material as food from its surrounding and excretes waste products to it. Thus the entropy of living organisms decreases, while that of their surrounding increases. Reversal of this process means death of the organism. This is the scientific way of expressing that creation is an ordering from initial chaos by pushing back chaos. Likewise, the scientific explanation of contingency is that created order, left to itself, always tends to revert to chaos (rather than to fall into "nothing"). One could also say that the second law has a spiritual parallel in the insight that closing ourselves to the Holy Spirit leads to spiritual death.

Another interesting insight is provided by *information theory*. Tom Stonier has derived a mathematical formula for the relation between information content and entropy of the cosmos.[39] This formula shows that at time zero entropy is infinite and information content zero, which would represent the initial chaos at the moment of the big bang. In the course of time information content rises steadily, while entropy decreases, indicating increasing order during evolution. Eventually information content approaches infinity and entropy goes to zero. This would represent the end of cosmic and biological evolution – in theological terms, the establishment of the New Kingdom. For cosmic evolution this would represent the turning point towards degeneration of the cosmos. For human evolution an approaching end is actually suggested by the low evolutionary rate of our species compared to animal species.[40] The rate for humans should eventually become zero due to our elimination of natural selection through the combined effects of medicine and technology. We are reminded here of Teilhard de Chardin's Omega Point,[41] which to him represents a mystical culmination of consciousness.

Chaos Theology

Contingency and improbability of the universe

As stated earlier, contingency means being dependent, endangered and accidental. The biblical account of creation implies contingency of the universe in the sense that creation comes about by the creative Word of God. If God were to withdraw his Word, the universe would revert to chaos. In the course of their development humans have become aware of this truth, which has led them to religious awareness. In early times this was a pre-scientific faith, but in our time it may be a faith enlightened and deepened through our scientific insight.

Science also recognizes the contingency of the cosmos, as I shall explain in some detail in sections 2.2 and 2.3. In cosmic evolution only the exquisite fine-tuning of the fundamental physical constants (e.g., strength of the fundamental forces, mass and charge of the elementary particles) has permitted the development of our universe and the formation of planet Earth on which life could arise and evolve. This means that we live in an extremely improbable, hence very contingent universe. In biological evolution contingency appears in the unpredictable and capricious course of evolution.

Our current Western society seems to be beset by anxiety – for the future of our world, for our inability to control rapid technological advances, for our very existence. Paul Tillich defines anxiety (theologians often use the German word *Angst*) as the awareness of the possibility of *non-being,* and concludes that this is the negation of every concept, the opposite of "being" (in the sense of "existing"), and that "being" has "non-being" in it.[42] Tillich's concept of "non-being" appears to be another term for the *nihil* of the *creatio ex nihilo* doctrine. However, a lapsing of existing things into "nothing" conflicts with our experience and with the physical law of conservation of mass and energy. Tillich's "non-being" can be replaced by the concept of "chaos" without any change in his conclusions. Anxiety then becomes the awareness of the ever-present danger of lapsing from ordered and meaningful existence into disordered and meaningless chaos; in short, anxiety is the awareness of the contingency of world and self. Interestingly, this fits very well with the claim of the psychotherapist-theologian

Eugen Drewermann that various types of mental illness are caused by the fear of being thrown back into primordial chaos, of which he sees a remaining element in our world.[43] In section 3.6 I argue that in physical illnesses like cancer there is a relapse into primordial chaos on the cellular level. All this indicates that the occurrence of contingency can be better explained by chaos theology than by *creatio ex nihilo*.

Progressiveness and purpose in evolution

Both believing and non-believing scientists must admit that science by its nature cannot deal with the concepts of progressiveness and purpose in the evolutionary process, which imply "intelligent design." When I see in this process the purposeful action of a loving Creator, I make a theological rather than a scientific statement. The same is true when I explain the "improbability" of our universe as the result of purposeful design by the Creator. As a scientist I should, however, always look for a scientific explanation as well as a theological interpretation.

The prominent evolutionary biologist Stephen Gould argues vehemently against recognizing any form of progressiveness in biological evolution. He says:

> Three billion years of unicellularity, followed by 5 million years of intense creativity and then capped by more than 500 million years of variation on set anatomical themes can scarcely be read as a predictable, inexorable or continuous trend toward progress or increasing complexity.[44]

Gould is speaking here about the so-called Cambrian explosion in evolution. If he had merely said that as a biologist he cannot use the concept of progress because it is a value judgment, then he would have been within his rights. However, every scientist must admit that mammals are more complex than bacteria, so by even denying increasing complexity in evolution, as he does, he exposes his real purpose: to deny the existence of a purposeful Creator. Nobelist-biologist Jacques Monod uses "chance" to argue against any form of purpose in evolution, but does so in a way that turns chance into a near-deity.[45] It is amusing to note that from the same scientific facts eminent, non-believing scientists, like Gould and Monod, can come to entirely different (supernatural) conclusions than those of believing scientists. This reminds me of the passage in Wisdom 13:1-9:

> For all people who were ignorant of God were foolish by nature; and they were unable from the good things that are seen to know the one who exists, nor did they recognize the artisan while paying heed to his works....

Even here, we might say, the remaining element of chaos is at work.

1.5 Chaos Theory and Chaos Events

What is chaos theory?

Relevant to chaos theology is chaos theory, the scientific theory of *chaos events*.[46] I must caution that the word "chaos" is used here with two entirely different meanings: "disorder" in chaos theology, "unpredictability" in chaos events. It is now recognized that many physical, chemical, and biological systems are governed by non-linear dynamic equations for their development in time. A typical example of such an equation is:

$$x_{n+1} = k.x_n(1-x_n).$$

It can be applied, for instance, to an insect population like a moth colony.[47] In that case n and n+1 indicate successive generations, and x the number of insects in a given generation. The proportionality factor k embodies the fertility level and other factors that remain unchanged for several generations. The number of insects is not only proportional to the size of the preceding generation (x_n), but also to $(1-x_n)$, namely food shortage and spread of disease in an oversized population. The factor k will also change in time, when there are alterations in fertility level or other factors. For a given number of moths in generation n, the number of moths in future generations may be calculated for changing values of k. This can best be done on the computer. Then we find a single line for values of k from 1 to 3.0, but then the line bifurcates, and at 3.45 both legs of the bifurcation bifurcate again, and so on. At k = 3.57 there are an infinite number of bifurcations; the system has become *chaotic.* The computer plot shows that, at a bifurcation, the population size may go up or down, but it does not show which leg will actually be followed, towards higher or lower values of x. The system has become unpredictable for us.

As there is no energy difference between the two legs, they have equal probability (just as for a bead at the top of a perfectly smooth wire in the shape of an inverted U there is equal probability of its descending along either leg). To the human observer the system has become fully unpredictable: not through lack of information on our part, but by being inherently unpredictable.[48] To those who claim that the system remains fully "deterministic" (i.e., it follows a natural law, in this case, a mathematical law), I say: Yes, the curve with the bifurcations is fully deterministic, but the moth colony can take only one of the two paths at a bifurcation. We cannot predict which leg it will take, since there is no energy difference between them. So the mathematical curve may be fully deterministic, but the natural system becomes indeterministic at a bifurcation point. A *chaos event* occurs.

Does this really happen in the natural world? First of all, we have to bear in mind that the equations represent models that approximate reality but cannot take into account all factors that may influence a natural system. Secondly, testing the model once the system has become chaotic is impossible: the computer plot describes all courses that the system may follow, while the natural system can only follow one course at a given time. Within these limitations, confirmation has been obtained in some tests, like those for a dripping faucet. The equations have also been found useful in the prediction of annual fluctuations in locust populations, a prediction which is helpful in combating such insect plagues.

Examples of chaos events

Chaotic systems possess flexibility and openness to novelty. An example is our solar system, which by computer simulation has been shown to be chaotic.[49] An uncertainty of only one kilometer in the present distance of 150 million kilometres between Sun and Earth increases to 150 million kilometres in the astronomically short time of 95 million years. This means that at this moment we cannot predict whether 95 million years from now the Earth will collide with the Sun or will be twice as far away from it as at present, or somewhere in between. Neither extreme may come about, but we cannot predict the distance between Sun and Earth after 95 million years from now. Another example is the weather: with all our progress in observation and computation, we are still unable to predict the weather for more than about five days ahead. Then it becomes unpredictable because of chaos events.

Chaos events may be operating in many cases of so-called "chance events" in cosmic and biological evolution. An example is the Cambrian evolutionary explosion, to which Gould refers in his quoted statement (section 1.4): after 3 billion years of evolution of unicellular life there arose in a mere 5 million years (beginning 540 million years ago) the ancestors of all currently existing life forms. Chaos events have, most likely, played a crucial role in this phenomenon. To the scientist chaos events make the universe unpredictable and open. The theologian may say that in these events God can act, if he chooses to do so, through the operation of the Holy Spirit without violating any physical law, because there is no energy difference between the legs of a bifurcation. This conclusion has previously been stated by John Polkinghorne.[50] Arthur Peacocke considers it at some length, but in conclusion says,

> "...it would seem that the unpredictabilities of non-linear systems do not as such help us on the problem of articulating more coherently and intelligibly how God interacts with the world. Nevertheless, recent insights of the natural sciences into the processes of the world, especially those on whole-part constraint in complex systems and on the unity of the human-brain-in-the-human-body, have provided not only a new context for the debate about how God might be conceived to interact with, and influence,

1: Chaos Theology as a New Approach

events in the world but have also afforded new conceptual resources for modelling it."[51]

He prefers to think that God acts on "the world-as-a-whole," but I find this rather vague idea not necessarily in contradiction with the suggestion that God may act in and through what we recognize as chaos events.

Can God act through chaos events?

Willem B. Drees thinks not.[52] First he says that if God were to act without input of energy, the action would be infinitely slow, as the Heisenberg relation for energy E and time t reads:

$$\Delta E \times \Delta t \geq h/2$$

(h is the Planck constant). So for $\Delta E=0$, the time t would become infinite. However, infinitely slow action is excluded, as the decisive input of information should take place before energetic disturbances have changed the situation. Secondly, he points out that information input always requires some energy expenditure, given by the equation

$$\Delta E \leq \Delta I \times k_B \times T \times \ln 2$$

(where ΔI is the information input, k_B the Boltzmann constant, T the absolute temperature in degrees Kelvin).

My calculations show that God faces no problem. The minimum information input required to influence a chaos event (+1 or -1) is $\Delta I = 1$, the Boltzmann constant k_B is 1.38×10^{-16} erg/°K, T may be set at 300°K, $\ln 2 = 0.6932$. This gives for the energy required

$$\Delta E = 3.3 \times 10^{-14} \text{ erg} = 0.69 \times 10^{-21} \text{ gcal.}$$

This energy would be withdrawn from the immediate environment, lowering its temperature by less than 10^{-21} °C. Substituting this calculated value of ΔE in the Heisenberg equation (Planck constant $h = 6.62 \times 10^{-27}$ erg.sec) gives:

$$\Delta t \text{ the value of } 1.0 \times 10^{-13} \text{ second.}$$

This is 10 to 100 times faster than the fastest chemical reaction. God passes with flying colours! Some may take objection to this kind of mathematical inquiring into the actions of God, but to them I suggest that as soon as one uses physical equations to make a theological point, then it behooves us to "make the sums," as the English say.

At the end of this discussion of chaos theory, I conclude that we may see here a convergence of the physical notion of "chaos events" and my theological notion of the "chaos element" in creation. I like to assume that the human spirit, in cooperation with God's Spirit, may also be operative in such events, e.g., in a medically unexplained remission of cancer (section 3.6). What the scientist observes as chaos events, the theologian and the believer may experience as the love of the Creator for his creatures.

1.6 Summary and Conclusions (Chapter 1)

Creation from an initial chaos, as described in Genesis 1 and 2 and in virtually all non-biblical creation stories, was replaced around AD 200 by *creatio ex nihilo* (creation from nothing) in the battle against Marcionite and Gnostic dualism. I show that a strict *nihil* presents serious problems of a conceptual, biblical, scientific, and theological nature, while a loose *nihil* as an "existing nothing" is not essentially different from an initial chaos. Thus I return to the biblical idea of creation from an initial chaos. This does not reintroduce Gnostic dualism, as long as we adhere with Genesis 1 to the one God who creates by his sovereign Word and we leave the initial chaos as a mystery, like the mystery of the big bang in cosmology. From this premise I develop a "chaos theology," which can clarify some important theological concepts and contribute to a reconciliation of the theological and scientific insights about the origin and destiny of cosmos and humankind.

Creation from chaos is described in Genesis 1 as separations and ordering of initial chaos by the Creator. In this process there remains an element of chaos, in the Bible often symbolized as the "sea," that, in the process of continuing creation will be abolished with the coming of a new heaven and earth (Revelation

21:1). This remaining chaos element, I propose, expresses itself in the evil existing in the world, the physical evil of natural disasters and disease and the moral evil committed by humans against each other and against nature. This resolves the perennial problem of the existence of evil in a world created by a good God (theodicy), a problem that is unsolvable in a *creatio ex nihilo* context. Jesus Christ, though fully human as well as divine, is unaffected by the chaos element. The incarnation marks the decisive beginning of the complete removal of the chaos element on the last Day.

The usefulness of this chaos theology for the reconciliation of the two world views is illustrated under several headings: initial mystery; separation and ordering; chaos and entropy; contingency and improbability of the universe; progressiveness and purpose in evolution. Finally I discuss the physical theory of chaos events, which explains how systems governed by non-linear equations (e.g., living beings) can become unpredictable. Such chaos events may operate in many phenomena in cosmic and biological evolution that used to be ascribed to "chance." I suggest that in continuing creation God may act by influencing chaos events through the Holy Spirit, without violation of physical laws, sometimes in cooperation with the human spirit.

Chapter 2

God's Action in the World

2.1 Influence of the Scientific World View

How does God act in the world? The answer to this question has been greatly determined by the dominant scientific world view of the time. The discovery of the laws of gravity and motion by Isaac Newton (1642–1727) led to a mechanistic world view. Once created, the universe would run a predictable course according to fixed laws. This led to a *deistic* view of the Creator, who, after one act of creation, left the world to develop by unalterable laws.

In the early 20th century, relativity theory and quantum theory with the uncertainty principle of Heisenberg replaced Newton's space-time determinism by space-time-matter-energy indeterminism, in the words of John Polkinghorne.[53] On the microlevel of elementary particles Newtonian determinism changes into quantum mechanical probability, Newtonian certainty into Heisenbergian uncertainty, waves behave like particles and vice versa. Space and time, observed as separate entities on the macrolevel (our daily world), merge on the galactic megalevel. Fundamental forces turn into particles at the megalevel of superenergetic particle collisions. Matter, forces, and energy are unified in a single microscopic fireball at the top megalevel of the big bang. Harold K. Schilling calls this the infinity and unfathomableness of the known physical reality, the mystery of ultimate reality.[54] In the late 20th century, another kind of unpredictability was discovered in the *chaos events* that occur in many physical, chemical, and biological systems (section 1.5). In the current scientific world view the universe is seen as open and spontaneous, a universe in which the theologian may see God's immanent and providential activity operating in chaos events.[55]

Another aspect, long recognized in theology and now strikingly evident in modern science, is the contingency of the universe, its being dependent, endangered and accidental. In sections 2.2 and 2.3 several examples of contingency in cosmic and biological evolution are presented. This leads to a discussion of the nature of God's action in the world in section 2.4 and of incarnation and reconciliation in the person of Jesus Christ in section 2.5, in which I apply chaos theology in combination with the theory of chaos events.

2.2 Contingency in Cosmic Evolution

It is now realized that cosmic evolution, which has eventually led to the appearance of the Earth and its inhabitants, has moved within very narrow tolerances. Gribbin and Rees speak of two *cosmic coincidences*:[56]

1. The universe is "tailor-made for humankind," meaning that the values of some 25 fundamental constants have just the right magnitude for cosmic evolution to have led to an environment suitable for the development of life and eventually humankind.[57] Consider these four examples:

(a) With a slightly weaker gravitational force the Earth would not have formed, but with a slightly stronger force the universe would have collapsed before life could have arisen;

(b) With a slightly weaker gravitational force, stars would have remained too cool for nuclear fusion to produce heavier elements (which are essential for the formation of planets and living beings), but with a slightly stronger force, the stars would burn up faster, thereby, in the case of the Sun, precluding biological evolution on Earth;

(c) If the strong nuclear force were 0.3% larger, no hydrogen could exist and the atoms essential for life would be unstable, but if it were 2% smaller, no elements heavier than hydrogen would exist, and thus no life; and

(d) If the weak nuclear force were slightly larger, too much helium would have been formed in the big bang and no heavy element ejection from stars would have occurred; if it were smaller, too little helium would have been formed, leading also to a lack of ejection of heavy elements from stars. Another astounding but inescapable fact is that the formation of this unbelievably large universe with its 30 billion galaxies and 3×10^{21} stars was needed to produce human life on at least one planet: a much smaller universe would have collapsed before advanced life could have arisen.

2. The universe appears to be "flat," meaning that our expanding universe is at the knife-edge between expanding forever or eventually collapsing in a fiery crash, the "big crunch." This

requires an extremely precise balance between the expanding force of the initial explosion and the gravitational force of better than one in 10^{60}, corresponding to the accuracy required to hit an inch-wide target on the other side of the observable universe.[58] A flat universe was predicted by the *inflation theory*, formulated by Alan Guth in 1981 to resolve certain problems in particle theory.[59] It assumes that 10^{-35} sec after the big bang a fast, brief expansion inflated the early universe from less than proton size to grapefruit size. Inflation theory and the flat universe have recently received further confirmation from stratospheric balloon measurements.[60]

Physical theory can note, but not explain, these cosmic coincidences. A kind of explanation has been attempted by Barrow and Tipler with their so-called *anthropic principles* (a.p.):

(a) the universe we observe must be compatible with our existence as observers (weak a.p.);

(b) the universe must have the properties which allow intelligent life to develop or else we would not exist (strong a.p.);

(c) the universe exists in a definite state, because it is observed by conscious beings (participatory a.p.); and

(d) the life that now exists in the universe will continue to evolve until reaching the point at which life is omnipotent, omnipresent, and omniscient (final a.p.).[61] To these explanations John Polkinghorne responds: the first one is just a tautology (a senseless repetition of meaning in different words), and the others are a form of teleology (belief in a purpose),[62] which is just what the drafters wanted to escape. It does not appear that the anthropic principles provide a satisfactory explanation for the "improbability" of our universe.

Another attempt has been made through some form of *multiworld hypothesis*. Richard Gott suggested that the big bang produced an infinite number of universes with different sets of fundamental constants, one of which happened to have the right set to lead to the development of the universe in which we live and the only one we can observe.[63] Lee Smolin claims that every time a black hole is formed, a new universe is spun off deep within it, each with a different set of constants.[64] However, this scenario appears to violate accepted astrophysical theory on several points.[65] A third form suggests two universes, one of which is ours, originating not from a big bang but from a "platelike splash."[66] It seems to me that science has so far failed to provide a rational and testable explanation for the improbability of our universe, and that it is more reasonable to believe in a purposeful Creator than in anthropic principles or multi-world hypotheses.

2.3 Contingency in Biological Evolution

Prebiotic evolution

For the origin of life on Earth from inorganic matter, as it currently is thought to have occurred, several examples of contingency can be formulated:[67]

(a) The initial oxygen-free Earth's atmosphere had the double advantage of allowing formation of biomolecules through solar ultraviolet radiation (absence of ozone layer) and at the same time protecting them from oxidation;

(b) Hydrothermal vents in the ocean bottom provided the right environment for the formation of the components of the first living cells;

(c) The "chicken-and-egg" problem of the present DNA -> RNA -> protein replicating system (for DNA to function certain protein enzymes are needed) has probably been solved by the finding that RNA has enzymatic properties, suggesting that RNA appeared first and provided a primitive replication system forming both DNA and various proteins (the "RNA-world");

(d) Water has some very propitious and unique characteristics essential for life: its expansion when freezing safeguards aquatic life, its ability to form hydrogen bonds allow DNA and protein molecules to adopt the right shape for their function, and its polarity allows the spontaneous formation of cell membranes from lipid molecules;

(e) After algae began to grow in the ocean their photosynthetic production of oxygen led to an oxygen-rich atmosphere with an ozone layer, shielding the emerging life on Earth from solar ultraviolet radiation;

(f) The formation of cells with a nucleus (eucaryotic cells) and oxidative metabolism led to plants and animals that could conquer the land; and

(g) The carbon cycle (uptake of carbon dioxide and release of oxygen by plants and the reverse by animals) maintains a constant 21% oxygen level in the atmosphere, high enough to permit the existence of large, advanced vertebrates and low enough to prevent spontaneous, large-scale forest fires.

There are other notable examples,[68] For instance, the distance of Earth to the Sun and its nearly circular orbit provide a suitable and sufficiently constant temperature on Earth. Actually the luminosity of the Sun increased by about 30% during the past 4 billion years, but algal growth and photosynthesis gradually lowered the atmospheric carbon dioxide level, so the surface temperature of the Earth has remained remarkably constant.[69] Additional examples include the Earth's strong magnetic field (800 times as strong as that of Mars), which diverts cosmic particle radiation, and the presence of massive Jupiter diverting asteroids from the Earth. A final example is the suitable surface gravity on Earth: if it were stronger, too much ammonia and methane would have been retained; if weaker, insufficient water would have been retained. These characteristics have given the Earth, as the only one of eight planets in our solar system, the right conditions for the development of primitive life and its evolution to intelligent life.

Biological evolution: questions and partial answers

The occurrence of biological evolution is supported by so much evidence from different fields of science (geology, paleontology, comparative anatomy and embryology, molecular genetics, radioactive dating) that there is no room for doubt anymore. The mechanism of evolution is now explained in terms of a succession of

2: God's Action in the World

mutation-selection steps. However, since experiments with processes of such long duration are difficult, there are still several unanswered questions, four of which I shall consider with the emerging answers.

1. How can new species arise when mating of members of different species usually leads to sterile offspring, like the mule resulting from mating of horse and donkey? However, fertile hybrids do occur in many plants and animals.[70] Moreover, when does the bar against successful mating with a mutant partner arise?

2. What about the lack of transitional forms in the fossil record? Transitional forms can be found by meticulous study of fossils.[71] Another explanation may be punctuated evolution (evolutionary steps alternating with long periods of standstill) for which evidence has been obtained in experiments with bacteria[72] and with guppies in nature.[73]

3. Why did creatures develop characteristics that had no apparent usefulness for survival? The tail feathers of the peacock have been cited, but these may provide a reproductive advantage. There may be no biological advantage in the replacement of asexual reproduction (splitting of a mother cell into two daughter cells as in bacteria) by sexual reproduction,[74] but there is the great social advantage of family life with a loving and caring relationship between male and female and their offspring.

4. How did organs of extreme perfection and complexity, like the eye, develop? It has been claimed that this would not have been possible by random mutation and selection.[75] However, in recent years several genes have been found that control the development of the embryos of widely different species.[76] For example, the pax-6 gene acts as a "regulator" in the development of all three eye types, the human eye, the "reverse" octopus eye, and the compound insect eye. When the pax-6 gene from the mouse is spliced into the fruitfly *Drosophila,* compound eyes, not mouse eyes are formed on the fly's wing.[77] This suggests that pax-6 stands at the top of three hierarchies of genes, each leading to one of the three eye types.[78] Several other genes, common to invertebrates and vertebrates and responsible for various aspects of embryonic

development, have now been found. Such genes may have originated in a common ancestor of invertebrates and vertebrates, an ancestor that was formed during the Cambrian explosion (see below).

Contingency in biological evolution

In biological evolution, contingency appears first of all in its unpredictable course. In 4 billion years of biological evolution billions of species have arisen, but only about one thousandth of these have survived. It has been a meandering process in which myriad possibilities seem to have been tried out. A major example of this unpredictability is the Cambrian evolutionary explosion (540 million years ago), which, after 3 billion years of sluggish evolution generating virtually only microbes and some algae, produced in a mere 5 million years all significant modern animal phyla. This evolutionary explosion has not yet been explained, but we must note that in the preceding 2 billion years photosynthetic algae were busily converting the initial oxygen-free atmosphere, necessary for the origin of life, into the present oxygen-rich atmosphere, necessary for the development of life.

Unpredictable catastrophes have played a decisive role in the course of the evolutionary process. At the end of the Permian period, about 250 million years ago, the greatest mass extinction in history took place: 80% of marine species and 70% of terrestrial species disappeared in a period of less then 165,000 years.[79] This catastrophe is ascribed to a series of massive volcanic eruptions in Siberia, possibly triggered by an asteroid impact, which caused global cooling and oxygen depletion in the single ocean of that time through a large discharge of carbon dioxide.[80] Another striking example is the extinction of the dinosaurs 65 million years ago, probably as the result of an asteroid impact.[81] It was disastrous for the dinosaurs, but opened the way for primate development and eventually for the appearance of humans.

These are examples of the contingency of biological evolution. In the complexity and unpredictability of the evolutionary process some see only evidence for a process without progress, plan or purpose, driven by chance (Monod) or by a "blind watch-

2: God's Action in the World

maker" (Dawkins).[82] To me it seems to point to the activity of a purposeful Creator, who allows his creation much freedom, but keeps it on course to his goal by occasionally influencing a chaos event.

Contingency in human evolution

All hominid species (*Homo habilis, erectus,* and *sapiens*) seem to originate from East Africa (Ethiopia, Kenya, Tanzania). This may be due to the Rift Valley tectonic event 8 million years ago, which turned East Africa into a dry, open savanna area without great forests.[83] This drove the primates out of the trees and onto the ground, leading to bipedalism in *Australopithecus,* the common ancestor of chimpanzees and humans. Bipedalism, which primarily arose from the need to spot predators on the dry, open savanna, led to important developments in the *Australopithecus* family. New activities of the hand (for defense and for gathering seeds and berries) had to be coordinated with the eye, causing the brain to grow in size and complexity. The freeing of the hands relieved the jaw from the task of grasping objects, thus permitting the development of throat, tongue, and mouth as later needed for speech. The larger brain size may also have led to living in families of an adult male with one or more females and their young for protection, as suggested by the finding of the "First Family" of thirteen *A. afarensis* individuals in Ethiopia. There were also negative effects of bipedalism: (a) the greater brain size required earlier birth, and thus the babies were more helpless than in earlier species and had to be carried in the arms of their mothers; and (b) the narrowing of the pelvis and the increased skull size made human childbirth painful and requiring assistance (the biological parallel of Genesis 3:16).

Further evidence of contingency in human evolution is shown by two recent molecular genetic studies. First, the present human races are genetically very close (varying only in 0.1% of their genome), in sharp contrast to the families of gorillas and chimpanzees. This appears to be due to the existence of a population bottleneck in *Homo sapiens* of some 10,000 individuals until 10,000 years ago. All this time the survival of humankind hung on a thread during the last glacial period. The end of this period and the advent of horticulture led to a population explosion and the present genetically very homogeneous human family of 6 billion.[84] This is a scientific confirmation of the fundamental unity of mankind, which is implied in the Genesis 1 creation story and expressed by Isaiah (Isaiah 49:6) and Paul (Galatians 3:26-29; Ephesians 2:11-22). Secondly, biological evolution in *Homo sapiens* appears to be coming to an end. The evolutionary rate (measured as changes per nucleotide per billion years of non-coding DNA sequences, which are thought to be insensitive to the pressure of natural selection) is very low (1.2) compared to that of monkeys (2.1) and rats (4.8).[85] Since we have eliminated natural selection for ourselves by the use of modern medicine and technology in coping with disease and environmental challenges, it appears that evolution of humans, but not of other species, may be ending or may have already ended.

2.4 Transcendent and Immanent Action of God

The abundant evidence for contingency in the evolution from the big bang to *Homo sapiens* (sections 2.2 and 2.3) appears to me to be the nearest thing to a proof of the existence of God that we as his creatures can approach. The belief in a divine Creator seems to be a more reasonable assumption than the "explanations" offered in anthropic principles, multi-world hypotheses, "chance" or a "blind watchmaker." In the introduction to this chapter (section 2.1) I mentioned the shift from Newtonian determinism to the indeterminism of quantum mechanical theory, unpredictable chaos events, and manifold contingency. This required also a shift from the belief in a *deistic* Creator to belief in a God who remains active in the world. In agreement with the biblical view, we must now see God as the Creator, who not only stands at the beginning, but who remains involved in his creation, protects it during its development, and guides it to the destination that he has determined from the beginning. How can we express this?

2: God's Action in the World

Continuing creation rather than conservation

Theophilus of Antioch (c. 185) described God's action in the world after the initial creation as "conservation," which idea rested more on the Greek Stoa than on biblical teaching. Calvin developed this idea from the assumption of an initially perfect creation, which subsequently was damaged by the fall of Adam. Conservation then becomes, so to speak, keeping the damaged vehicle running until, at the last Day, it will be fully repaired or replaced by a new one. In the meantime God is more or less absent from the world. In my opinion the doctrine of conservation degrades the greatness of the Creator and overrates the power of humans. Instead, in chaos theology I hold to the idea of a continuing creation. As mentioned in section 1.3, the Hebrew word *tov* for "good" in the phrase "And God saw that it was good" means that the initial creation is good for its purpose, the goal that God has in mind in his creative work. In the continuing creation God is battling the remaining element of chaos, which is the source of the physical and moral evil in our world.

How does God interact with his creation?

To this question the volume *Chaos and Complexity* was devoted.[86] Even though the relevance of chaos events was the primary topic of discussion, three authors, Thomas Tracy,[87] Nancey Murphy,[88] and George Ellis,[89] still believe that quantum events are the vehicle for God's interaction. Both John Polkinghorne and Arthur Peacocke argue against this idea. Polkinghorne notes that quantum events play at the micro-level of elementary particles and atoms (e.g., we can know the average time elapsing between decays of the atoms of a radioactive element, but we cannot tell when a particular atom will decay), but it is difficult to see how such quantum events at the micro-level could be amplified to the macro-level of our daily existence.[90] Peacocke writes: "The unpredictability of quantum events at the sub-atomic level are usually either ironed out in the statistical certainties of the behavior of large populations of small entities or can be neglected because of the size of the entities involved."[91] Neither can there occur what has been called "quantum chaos." The fundamental equation in quantum theory, the Schrödinger equation, is linear, and thus it cannot lead to chaotic behaviour, as the non-linear equations of chaos theory do. These considerations dispose, in my opinion, of the relevance of quantum events for God's interaction with his creation.

For chaos events, which can involve complete living beings and even the entire solar system, there exists no amplification problem. As shown in section 1.5, God can influence a chaos event by information input speedily and with negligibly small energy expenditure, or, in theological terminology, by the Holy Spirit. This opens a way for God's action in the world without violation of physical laws and for his actions in human beings without violation of their free will. It is not necessary to assume that God will interact in every chaos event, as he leaves a great degree of freedom to his evolving creation.

Transcendent and immanent

This leads me to the view that God is acting in two ways: transcendently and predictably; immanently and unpredictably. Predictably God works through the natural laws, which are the human formulation of the orderliness of natural events as ordained by God in the initial creation. In accordance with these laws God creates and assures a reliable existence for all his creatures. Here we see God outside and above his creation, as *transcendent*. Unpredictably for us, God works in his creative freedom through the influencing of "chaos events," by which he guides his creation to the destiny determined by him in the beginning. Within the ordered structure described by natural laws God retains freedom for himself in his creative use of "chaos events." Here we see God as active within his creation, an activity for which I use the word *immanent*. God's immanent activity is invisible to us, except in hindsight in the course of evolution and in the course of our own life.

This dual activity reflects the loving providence of a God who acts both transcendently and immanently, and to whom petition-

2: God's Action in the World

ary prayers can sensibly be addressed. As Bishop Hugh Montefiore writes:

> ...the immanence of God in his creation... is wholly consonant with the traditional doctrine of God, and it seems to be consonant also with the phenomena of the evolutionary process investigated in the natural sciences. ...God is not only the creator, upholder and sustainer of all that is, but all being participates in his being, and his glory is to be seen in the smallest particle of matter as well as in its evolving forms.[92]

He quotes Aubrey Moore, who said, "Either God is everywhere present in nature, or he is nowhere. In nature everything must be his work or nothing."[93] This is beautifully expressed in the words of the psalmist in Psalm 104.

With regard to God's interaction with the world with its unpredictability, open-endedness, and flexibility, Arthur Peacocke considers that God in his omniscience knows all relevant details, and so knows all possible outcomes. This means that God can also manipulate the system to produce the desired outcome. But then he takes a step back, saying, God would then be "intervening in the order of nature with all the problems that that evokes for a rationally coherent belief in God as the creator of that order." This leads him to the conclusion: "It would seem that the unpredictabilities of non-linear dynamic systems do not as such help us in the problem of articulating more coherently and intelligibly how God interacts with the world."[94] It would seem to me that Peacocke's negative conclusion is the inevitable result of his rejection of God's "intervening" in the world. What, I ask, is wrong in assuming that God – who is immanently involved in his creation and who laid down in the initial creation all physical laws, including those of chaotic behaviour – will utilize chaos events for his ultimate creative purpose? If we deny God's "intervention" in his creation, we are back to the deistic God of Newtonian thinking.

Other terms for God

John Polkinghorne speaks about the *God of being* (in the timeless regularity of physical law) *and becoming* (in the evolving history of complex systems in chaos events).[95] I consider these qualifications inadequate: "being" is too passive a term in my view for God's transcendent activity, while "becoming" sounds too much like process theology.[96] The term *panentheism* is used by some to express that the world is in God, yet God is ultimately other than the world.[97] I consider this term unsuitable, because (a) it is easily confused with *pantheism,* where God becomes one with the universe, as seems to be the case for the god of process theology who develops with and in the evolutionary process, and (b) it does not express clearly the two different modes of God's activity in the world as perceived by us.

So I prefer to use the terms "transcendent" and "immanent." The immanence of God in his creation means that he is not merely "conserving" his creation, but remains creatively active in this world in the continuing creation, in his struggle with the remaining chaos element until its definitive removal on the last Day. I reserve the word "conservation" for God's active upholding of the contingent universe, without which it will revert to chaos. We may compare this with the action of the Sun's radiation on Earth as it counteracts and overcomes the inexorable tendency of all natural systems towards increasing disorder (entropy) and death. So God is creating and conserving simultaneously. As Montefiore says, "In this way he is leading the development of the universe to greater complexities until conscious mind emerges in human beings."[98]

Implications for prayer

God's transcendence in instituting physical laws implies that in prayer we cannot properly ask for a change of the seasons, because God has fixed these in the laws governing the movements of the Earth in the solar system. On the other hand, his immanent activity in influencing chaos events means that the weather and our bodies can be a proper subject for prayer, as they are prone to chaos events. In these events God may act through his Holy Spirit without violating physical laws (see section 1.5). We may trust that God in his perfect wisdom will know whether it is in the long-term interest of ourselves and all his creatures to grant our petitions. A miracle can be understood as a manifestation of God's immanent activity in chaos events, but not as a deviation from physical law, a deviation which would probably have disastrous consequences. In theological terms, a miracle is a sign of God's immanent love and power.

On the other hand, I have suggested (section 1.5) that, besides the Holy Spirit, the human mind may also be operative in a chaos event, as for instance in a medically unexplained remission of cancer (see section 3.6). Here I see a cooperation between God and human, between Spirit and mind, as becomes clear from Jesus' frequent call on the faith of the sick person in his healing acts. However, to me the greatest miracle is God's immanent activity in his continuing creation, in his ongoing battle with remaining chaos, culminating in the life, death, and resurrection of Jesus Christ, offering us reconciliation and the assurance of the New Kingdom. With Austin Farrer we may say that God's Spirit lives and works in the world, as the human mind in the human body, but with God granting a great degree of autonomy and freedom to the world, as the human mind does to the body.[99]

Problems with a transcendent-only image of God

The transcendent-only image has dominated in the western Church, not only among the deists, but also in Calvin's doctrine of conservation. In the latter doctrine creation is seen as immediately complete and perfect; there is only an initial creation and no continuing creation. The imperfection, noticed as the evil in our world, is attributed to the Fall, the primordial sin of Adam and Eve. Thereafter God has to "repair" his creation and he does so through Jesus Christ. My theological objection is that this overrates human power (in damaging a good creation by human sin) and underrates God's power by seeing him as a "repairman." The scientific objection is that physical evil in the form of natural disasters and disease was present long before humans came to the scene, while humans have begun to affect creation only in Earth's biosphere and only noticeably during the last two centuries (ecological crisis).

Problems with an immanent-only image of God

The immanent-only view of God's activity is a key feature of Whitehead's *process theology,* in which God is seen as developing in time, as realizing himself in and through the evolution of his creation.[100] Whitehead came to this idea, partly in protest against the "cosmic tyrant" of Calvinist theology, partly from the desire to integrate evolutionary theory in theology. This led him to the following four statements:

(a) life is found in the choice of novelty and in the quest for aesthetic enrichment;

(b) God acts in this process as a localized "lure" opposing the move towards increasing entropy;

(c) change is not by chance, but is an expression of a subjective aim at an unconscious level; and

(d) God is limited by allowing self-determination to the evolving universe.[101] Serious criticism can be levelled at this approach. John Polkinghorne says, "The God of Whitehead is a curiously passive deity."[102] Keith Ward[103] writes, Whitehead has changed the "cosmic tyrant," against whom he so strongly protests, into a "cosmic sponge." A god who develops with and in his creation, and who has to "lure" his creatures into evolving is not the biblical God of the universe. The God of the Bible is involved in the world, but not tied to it to the extent that the process theologians suppose. Process theology also leaves unanswered the questions How does the "lure" work, particularly at the lower levels, where physical laws operate rather than human free will? How can what is said about the becoming of a human person apply to a nucleotide? Where do human sin and its consequences enter? The immanent-only view of process theology does not offer a valid answer to the question of God's action in the world. It seems clear to me that we cannot do without a dual image of God, that of God as both transcendent and immanent.

2.5 Jesus Christ and Incarnation

The most central tenet of the Christian faith is that Jesus Christ is the incarnation of God in the human person of Jesus of Nazareth. What can chaos theology contribute to this tenet? In the New Testament we find two descriptions of the incarnation: the virgin birth story in the gospels of Matthew (1:16-25) and Luke (1:26-38) and the incarnation of the Word in the prologue to the gospel of John (John 1:1-14). The virgin birth story tells us that Mary was still a virgin when she was made pregnant by the Holy Spirit. Current biological knowledge informs us that for the conception of a human being the union of an ovum from the mother and a sperm cell from the father is required. The ovum contributes the X-chromosome, while the sperm cell must contribute the Y-chromosome to produce a male child. So the virgin birth is a biological impossibility, unless we assume that God miraculously created in the womb of Mary the Y-chromosome or the entire fertilized ovum.[104] However, this assumption poses a serious theological problem: Jesus would then not be fully human, and this would make him irrelevant for our salvation.[105] Thus, we can better leave the virgin birth as a pious legend.

This problem is overcome in the Johannine idea of the incarnation of the *Logos,* the creative Word of God, in the fully human Jesus of Nazareth. Jesus thereby became fully divine while remaining fully human. This safeguards the doctrine of the incarnation, which is far more crucial than the story of the virgin birth. John Robinson notes that John 1:1 implies neither an identity between God and Christ, nor a simple likeness between them.[106] The Greek version reads: *kai theos en ho logos.* Usually this has been translated, "The Word was God," which suggests identity. But, says Robinson (who was a New Testament scholar), if that is the case, the Greek would have placed the article with God: *ho theos.* Neither is John saying that Jesus is a "divine" human, in the sense of the Greeks, because then *theios* would have been used instead of *theos.* Robinson translates instead, "What God was, the Word was" (as in the NEB), which is not an identity, but suggestive of a common quality or substance. In Jesus we *see* God (John 14:9). He was the complete expression of God, which is how Phillips translated John 1:1: "At the beginning God expressed himself. That personal expression, that word, was with God, and was God."[107]

We may couple this with the idea of the corporate Christ[108] with the reciprocal indwelling of Jesus in the Father, the Father in Jesus, and we as believers in him and the Father: "As you, Father, are in me, and I am in you, may they also be in us...." (John 17:21, and many other texts from John and Paul). Then we obtain the full meaning of Christ as the mediator between God and man.

Back to the question posed at the beginning of this section: Can chaos theology contribute anything to the topic of the incarnation? If anything, I think, it can give us a deeper insight into God's purpose in the incarnation. Over the centuries popular Christian belief has narrowed down the significance of the incarnation to being merely the prelude to *our* salvation. However, Paul already recognized in Jesus the cosmic Christ, when he writes, "in Christ God was reconciling the world [Greek kosmos] to himself" (2 Corinthians 5:19) and "through him God was pleased to reconcile to himself all things, whether on earth or in heaven" (Colossians 1:20). And John writes, "that the world [Greek *kosmos*] might be saved through him" (John 3:17). Jesus as the cosmic Christ fits with our knowledge of cosmic evolution. The hydrogen resulting from the big bang condensed into stars, which through nuclear fusion produced the heavier chemical elements. After the exhaustion of the nuclear fuel these stars turned into supernovae, which exploded and ejected these elements as "cosmic dust" into the interstellar space. Eventually, in one galaxy, Sun and Earth were formed through condensation and accretion of these elements from a cosmic dust cloud. In prebiotic evolution, living cells were formed from these elements, and in the biological evolution all living beings, including ourselves, have been formed from these elements through the uptake of food. In this way, we humans have part in, are united with, the entire cosmos. Jesus, being fully human, also shares in this cosmic union, and thus through the incarnation he becomes the cosmic Christ.

In chaos theology I adopt on biblical grounds the idea of an initial incomplete creation, which God through his creative Word brings in the continuing creation to completion through abolishing the remaining element of chaos. It is in this process that, in the fullness of time, 15 billion years after the beginning of time in the big bang, God incarnated his Word in Jesus of Nazareth, who as the cosmic Christ is bringing the entire cosmos, including all humans, to completion and fulfillment on the last day. This is the full picture of reconciliation, of which our reconciliation is a part.

2.6 Jesus Christ and Reconciliation

What does reconciliation mean and how is it brought about? Dutch Calvinist theologian C.J. den Heyer caused quite a stir in his church a few years ago with his book *Verzoening* [Reconciliation]. After an extensive review of New Testament teaching on this topic, he admits that he can no longer accept the traditional Calvinist teaching about reconciliation.[109] He objects to the idea that God could be so entrapped in his own justice that blood must flow to achieve reconciliation, but den Heyer does not offer an alternative. Sadly, he has had to resign from his position as a professor of New Testament theology. Here I shall attempt to show how chaos theology can provide a better interpretation of the biblical message.

Reconciliation in the Bible

The Old Testament tells us about the covenant that the LORD made with Moses and his people, and about the Law he gave them. Anyone who fully and steadily kept the Law was assured of God's acceptance. But over the centuries a question arose as to how God could forgive the transgressor of his Law without compromising his own perfect justice. The prophets came to see that this is impossible. Jeremiah speaks about a new covenant (Jeremiah 31:31-34), but without solving the problem that for the Israelites blood was required for reconciliation; the blood of oxen fulfilled the requirement in the first covenant, but whose blood would be required in the new covenant? Isaiah suggests a solution when he writes about the Suffering Servant: "he was wounded for our transgressions; he shall bear their iniquities" (Isaiah 53:5, 11). The Old

Testament prophets also realized that without repentance the offering of sacrifice for sins was futile: "I have had enough of burnt-offerings of rams.... I do not delight in the blood of bulls..." (Isaiah 1:11); "I desire steadfast love and not sacrifice, the knowledge of God rather than burnt offerings" (Hosea 6:6).

In the New Testament Jesus begins his ministry with a call to repentance for sin, saying, "The kingdom of God has come near; repent, and believe in the good news" (Mark 1:15). He affirms the uselessness of animal sacrifices as a substitute for repentance, repeating the words of Hosea: "I desire mercy, not sacrifice" (Matthew 9:13). The "ransom" image (a metaphor of the slave market) appears in his words "the Son of man came...to give his life a ransom for many" (Mark 10:45). At the institution of the Eucharist he answers Jeremiah's problem by declaring, "This is my blood of the (new) covenant, which is poured out for many for the forgiveness of sins" (Matthew 26:28). He also applies to himself the words of Isaiah 53 about the Suffering Servant: "For I tell you, this scripture must be fulfilled in me" (Luke 22:37). When John the Baptist describes Jesus as "the Lamb of God who takes away the sin of the world" (John 1:29), Jesus' death is compared to that of the paschal lamb at the Passover.

Paul proclaims that "Christ died for our sins" (1 Corinthians 15:3). For him Christ's death and resurrection are the means by which we are redeemed from the effects of our transgression of the Law, from God's condemnation, and from (spiritual) death. By baptism we share mystically in Christ's death and his victory over it in the resurrection, and acquire justification as God's free gift. Peace was made between God and humans through the blood of the cross (Colossians 1:20). Paul also uses a "substitution" image: "For our sake he made him to be sin" (2 Corinthians 5:21). Once Paul uses the image of the paschal lamb – "our paschal lamb, Christ, has been sacrificed" (1 Corinthians 5:7) – although he knew very well that the lamb was only killed in the temple, but then taken home to be eaten at the *seder* meal. In Hebrews we find the image of the high priest who offered himself up (Hebrews 7:27). The Bible tells the story of our redemption in various images and metaphors.

2: God's Action in the World

Subsequent theological elaboration

Through the centuries theologians have attempted to formulate a theology of reconciliation from these biblical images. The following are some of the main examples with a brief characterization:

1. *Origen* (*c.* 200) formulated the *ransom* theory: The death of Christ is the ransom that had to be paid to Satan, who had acquired rights over all humans by Adam's fall.

2. *Augustine* (*c.* 400) accepted the ransom theory and connected reconciliation with the idea of original sin, the "hereditary" disease of humankind after Adam's fall (section 3.3).

3. *Anselm* (*c.* 1100) developed in *Cur Deus Homo* the *satisfaction* theory: As finite humans cannot make satisfaction to the infinite One, God himself had to take their place in the person of Christ, and by his death to make complete satisfaction to his divine justice.

4. *Thomas Aquinas* (*c.* 1250) denied the necessity of this method of satisfaction, but stated that the manner which God chose was highly fitting (congruous): it shows forth his omnipotence, goodness, wisdom, and the harmony between his justice and mercy. Reconciliation to him is the free gift of God to us who cannot redeem ourselves due to the loss of the supernatural gifts by original sin. Humans may co-operate (by faith *and* works) with God's grace in receiving justification and sanctification. This became the traditional Catholic teaching on reconciliation.

5. *Luther* vigorously denied the possibility of human co-operation with grace other than by faith alone *(sola fide)*. Claiming to return to Pauline teaching, he rejected the satisfaction theory and replaced it with the *substitution* theory: Christ, bearing by voluntary substitution the punishment due to us, was reckoned by God a sinner in our place.

6. *Calvin* went even further with his penal theory: Christ bore in his soul the tortures of a condemned and ruined man.

From all these theories emerges, in my opinion, a gruesome image of a god who is so hopelessly imprisoned in his own justice that he must sacrifice his own Son for the reconciliation of humankind. Like den Heyer, I cannot accept such doctrines. In addition, I object to the anthropocentric view of reconciliation, which completely neglects the need of the entire cosmos for reconciliation and fulfillment. In my opinion, theologians from Origen to Calvin have made three errors:

(a) they literalized biblical metaphors;

(b) they isolated the death of Christ from his incarnation and resurrection; and

(c) they neglected the continuing creation as God's ongoing battle against evil stemming from remaining chaos. As a witty illustration of the danger of literalizing metaphors I add the following statement by my friend the Rev. Dr. J. Verburg: "A sign with the symbol for 'exit' is not itself the exit, but points to it; whoever thinks that it is the real exit will bump his head against it and not get out."

Reconciliation in chaos theology

Chaos theology (section 1.3) offers in my opinion a better understanding of reconciliation through Jesus Christ. God, in his continuing creation, has been involved in an ongoing battle with remaining chaos for 15 billion years of our time already, in which time humans have existed only during the last 10,000 years. So this battle is not only a human predicament, but a cosmic drama. Paul senses this when he says, *"the whole creation has been groaning in labour pains until now"* in expectation of liberation (Romans 8:22). Therefore, in this ongoing battle God is not redeeming only humans, but the entire cosmos. In section 2.5 I concluded from the words of Paul (2 Corinthians 5:19; Colossians 1:20) and John (John 3:17) that Jesus is the cosmic Christ, who is reconciling the entire cosmos to its creator. This view is much more satisfactory in my opinion than the anthropocentric way in which traditional theology has treated reconciliation, considering only the reconciliation of humans.

Of course, it is not a trivial matter that God's top creatures, his image bearers, succumb again and again to the remaining chaos element and thus become sinners. Remaining chaos, in what one might anthropomorphically call "a last desperate effort," leads humans to kill Jesus in the crucifixion, as many martyrs have been killed in the course of history. However, God turns the apparent defeat by remaining chaos into victory by the resurrection of Christ. This is an initial victory, which will become definitive on the last Day, when God will forever banish the chaos element. It is the total action of Christ, rather than merely his death, which brings reconciliation to the cosmos and to us. All humans, being the only creatures (as far as we know) who have knowledge of sin and have freedom to withstand it, may receive the benefits of this process in faith (Romans 4:1-16). This theology of reconciliation avoids literalizing the biblical metaphors, integrates the crucifixion with Christ's incarnation and resurrection, gives reconciliation a cosmic dimension, and places it in God's continuing creation leading to the fulfillment on the last Day. God is not pictured as a captive of his own justice. Crucial is our acceptance, in and through faith, of the reconciliation achieved in Jesus Christ; only then can we become inhabitants of the New Kingdom, which is creation fulfilled.

2.7 Summary and Conclusions (Chapter 2)

God's action in the world is considered in the light of the contingency of the creation. In the current scientific world view with quantum indeterminism and unpredictable chaos events, it has become clear that the universe is contingent, i.e., dependent, endangered and accidental. Examples in cosmic evolution are the two cosmic coincidences: (a) the universe could only develop as it did, owing to a fine-tuned set of values of 25 fundamental constants; and (b) the universe seems to be on the knife edge between forever flying apart and ultimate collapse. Science cannot explain these cosmic coincidences. Two other points are noted. First, this vast universe with its billions of galaxies and stars was needed to permit humans to arise on Earth. Finally, the second law of thermodynamics tells us that the universe, if left to itself, would revert to chaos.

Contingency in biological evolution is visible in the initially oxygen-free atmosphere and the presence of hydrothermal vents in the ocean bottom that permitted the origin of life from inorganic matter on Earth. Photosynthetic algae then produced an oxygen-rich atmosphere with an ozone layer, leading to plant and animal life on land. Contingency is also noticed in the Cambrian explosion, when, after 3.5 billion years of prebiotic and microbial evolution, virtually all body plans of present living beings developed in a mere 5 million years. The extinction of the dinosaurs by an asteroid impact 65 million years ago allowed development of primates and eventually of humans. The unpredictability of biological evolution suggests the operation of chaos events.

Contingency in human evolution appears in the Rift Valley tectonic event 8 million years ago, which turned East Africa into a dry, open savanna area with few trees, making it the birthing ground of all hominids. A population bottleneck in *Homo sapiens* of some 10,000 individuals until 10,000 years ago produced the present, genetically very homogeneous, human family of 6 billion people, confirming the biblical idea of the unity of mankind. Evolution of *Homo sapiens,* but not that of other species, appears to be coming to an end, suggesting that the destiny of the creation process has been approached.

The biblical idea of a purposeful Creator offers a more likely explanation for the multiple contingency of creation than is offered by anthropic principles, multi-world hypotheses or chance. The Creator acts both transcendently in the initial creation and immanently by remaining involved in his creation, protecting it during its development, and guiding it in his ongoing battle with remaining chaos to the destiny he chose for it. A one-sided image of God, either transcendent or immanent, encounters theological as well as scientific problems.

The work of Jesus Christ is considered under the headings of Incarnation and Reconciliation in the light of chaos theology. The virgin birth story of Matthew and Luke, in the light of current

2: God's Action in the World

biological knowledge, presents the problem that such a conception would render Jesus not fully human and thus irrelevant for our salvation. John's account of the incarnation of God's creative Word avoids this problem, presenting Christ as fully human and fully divine. From biblical evidence and our insight into cosmic evolution, we may conclude that Jesus is the cosmic Christ, who is in union with the entire cosmos and who will bring the entire cosmos, including all humans, to completion and fulfillment on the last day.

Traditional teaching about reconciliation is faulted for three errors: biblical metaphors are literalized, the crucifixion is isolated from incarnation and resurrection, and continuing creation is neglected. This portrays God as a prisoner of his own justice, and therefore must sacrifice his Son to provide reconciliation. Instead, I start from God's ongoing cosmic battle against remaining chaos in continuing creation. The crucifixion is seen as "a last desperate effort" of remaining chaos leading humans to kill Jesus. God turns this apparent defeat into decisive victory through the resurrection of Christ. This promises definitive victory on the last Day with the banishing of the chaos element and the establishment of the New Kingdom including all creatures. Humans, as the only creatures who have knowledge of sin and have freedom to act, can become inhabitants only by accepting, in and through faith, the reconciliation achieved in Jesus Christ.

Chapter 3

The Problem of Evil

3.1 Evil: Prominent Topic of Discussion

Through our modern means of mass communication, our awareness of the problem of evil encompasses larger and more terrifying dimensions than ever before. There is the moral evil of horrible atrocities, captured under names like Auschwitz, Rwanda, Bosnia, Kosovo, Sierra Leone, and the World Trade Center (September 11, 2001). Nearer home we see unprovoked violence in our streets and homes, fraud and corruption by public figures. There is the physical evil of natural disasters in the form of earthquakes, volcanic eruptions, storms, and floods, by which thousands of innocent people are killed, injured, or made homeless. Another form of physical evil we face is the illnesses of ourselves or loved ones.

This cannot fail to raise the questions Why? Whence this evil? It is remarkable that persons who have distanced themselves from a belief in a personal God are still attributing the responsibility for the existence of evil to "God." It seems to be their final thought about the God who is disappearing from their view. A columnist in a prominent secular Dutch newspaper asks,

> If God has created the world, why is there then so much misery? And if he cannot do anything about this, why is he then called almighty? And if he is almighty, but does not intervene, then he surely is not good? Isn't it time to realize that the God of the Bible is an impossible construction? That whoever believes in this is fooling himself?[110]

She is repeating the questions already asked in 300 BC by the Greek philosopher Epicurus.[111] Many of us seem to have the simplistic idea that if God is good and omnipotent, he should not let bad things happen to us. This may explain why Rabbi Harold Kushner's book *When Bad Things Happen to Good People* became a bestseller.[112] I shall return to his answer in section 3.4.

Job is the eternal symbol of a good person who suffers all kinds of physical evil, until he has nothing left but his body covered with sores. The introduction of Satan as the agent inflicting evil on Job does not absolve God, because in the story Satan is authorized by God to test Job by causing all this evil to him (Job 1:12). On the other hand, the attempt to absolve God from responsibility led Marcion (*c.* 150) and the Gnostics to the idea of creation from evil matter by an evil demiurge. This in turn led Theophilus of Antioch (*c.* 185) to the invention of the doctrine of *creatio ex nihilo* (section 1.1), but this implies that God creates everything and thus also evil. This problem has never been satisfactorily solved by Christian theologians in the context of *creatio ex nihilo*.[113] Even Pope John Paul II seems to admit this in his encyclical *Fides et Ratio*.[114]

How then shall we explain the existence of evil in God's creation without lapsing into dualism or limiting either God's goodness or his omnipotence? In answering this question we must distinguish between moral evil (the evil that humans commit to each other, to nature, and to God) and physical evil (natural disasters and disease). On occasion the former may be involved in the latter, e.g., the loss of life in a flood, an earthquake or an airplane crash may be partly or wholly due to human negligence (lacking maintenance or repair) or greed (when proper measures are omitted for financial gain). In the following sections I shall discuss moral evil and human ambivalence (section 3.2), the problem of original sin and predestination (section 3.3), evil in the context of *creatio ex nihilo*, (section 3.4), evil in the context of chaos theology (section 3.5), and finally, disease as a form of physical evil (section 3.6).

3.2 Moral Evil and Human Ambivalence

The story of human origins in the first three chapters of Genesis provides us with great insight into the mysterious fact of human ambivalence. Humans are created in the image of God, who blesses them and tells them to rule over all the earth and its creatures (Genesis 1:26-30). They are thus given the pre-eminent place in creation. A sharp distinction is made between humans and all other creatures: (a) Only humans have the likeness of God (Genesis 1:26-31), and are *image-bearers* of God (i.e., only humans can communicate with God, can reflect about themselves, their surroundings and their Creator, and can know the difference between good and evil); and (b) they are given dominion over all the Earth and its creatures. Yet, being created means that humans, like all of nature, have their deepest ground of existence not in themselves, but in God, their Creator. They are earth-bound creatures, dependent on God's life-giving spirit. This means that humans are finite and limited. This finiteness is neither sinful nor shameful, but a normal condition of creatureliness. Being made in the image of God implies that humans are to use their God-given talent of creativity to cultivate the Earth in the interest of all animate and inanimate creatures. They are to be *co-creators* with God, and their culture, science, and technology are in principle God-given and God-willed. To rule over the Earth and its creatures means that God makes us his stewards, who are to make and keep the earth habitable, to populate and cultivate it, to care for animals and plants, and to make use of them in a caring way. Being God's image-bearers and co-creators is human greatness.

However, humans are also "fallen" creatures. In Genesis 2:8-17 we read how God placed the proto-humans Adam and Eve in the garden of Eden, an oasis in the desert, the desert being the initial chaos from which God creates Adam as well as the garden (Genesis 2:5-7). Here the garden stands for the state of unity and harmony with God in which humans were destined to live. God tells them to cultivate the garden and to feel free to eat from any of the trees, except from the tree of the knowledge of good and evil. "Knowledge of good and evil" is a Hebrew expression for the knowledge of everything, the full and comprehensive knowledge that brings to its possessor the power and independence that rightfully belong only to the Creator. Such power and independence humans cannot properly handle by themselves, as shown over and over again in the history of human technological development. Yet, they are continually tempted to acquire this power. In Genesis 3:1-24 we read how Adam and Eve succumb to the temptation to eat from the fruit of this tree. Suddenly, they see their nakedness, the scales fall from their eyes. They see themselves as they really are, in their spiritual and moral nakedness, as fallen and mortal creatures, who are not satisfied with being God's image-bearers and stewards over the entire created world, but who want to be *equal* to God. Thereby they cut themselves off from their nearness to God in the ordered creation he gave them, and they become wanderers in the chaos of human life. This mythical story is symbolic for the condition of humans, who again and again grasp for equality with God. This is human *brokenness.*

Humans are, therefore, *ambivalent creatures,* embodying two opposing traits in one creature: greatness and brokenness. Their greatness is being God's image-bearers who may rule over the entire earth and its creatures, and who may be co-creators with God in the pursuit of culture, science and technology for the well-being of humanity and the world, and to the glory of God. Their brokenness makes them restless aspirers whose perpetual grasping for equality with God is the root of every kind of human evil. The murderer usurps the divine power over the life and death of his fellow. The thief, whether a petty thief or a million-dollar embezzler, violates God's sole ownership of all created things, over which we are merely to be stewards. This is true for acts committed by individuals as well as by nations. We also grasp for equality with God when we misuse our God-given creativity, our science and technology, for our own glory, power, and wealth. Human ambivalence taints all our activities, including our science and technology, so that every application of new scientific knowledge brings with it its own problems and potential for evil.

The Genesis accounts of human ambivalence may offer a *description* for the occurrence of moral evil, but it does not offer an *explanation* of the phenomenon. Why do creatures at the top of the hierarchy of creation behave in this way; how could God allow this creationary "mishap" to occur? So the problem of moral evil is part of the total problem of evil, which will be further discussed in sections 3.4 and 3.5 after first dealing with the doctrines of original sin and predestination (section 3.3).

3.3 Original Sin and Predestination

Original sin

The doctrine of original sin was first formulated by Irenaeus (c. 190) in his struggle against Gnostic dualism. Since he had accepted the *creatio ex nihilo* idea to combat gnostic dualism, he had to find an origin for sin outside God. This he found in Paul's words: "As sin came into the world through one man [i.e., Adam]...many died through the one man's trespass" (Romans 5:12-21). He interpreted these words to mean that evil came into the world through the sin of Adam. Didymus of Alexandria (*c.* 350) taught that Adam's sin was transmitted by natural propagation, and Chrysostom (*c.* 390) attributed this to sexual lust. Theodore of Mopsuestia (*c.* 400) was the only Greek Father of the first four centuries who explicitly rejected the idea of original sin and placed sin entirely in the human will. Augustine (*c.* 400) affirmed Chrysostom's idea about sexual lust as the driving force, claiming that Adam's guilt turned humanity into a *massa damnata*. The Augustinian doctrine was confirmed by several Councils, particularly the Second Council of Orange (529).

The medieval theologians attempted to define the nature and transmission of original sin. Anselm of Canterbury (*c.* 1080) defined it as the "privation of the righteousness which every man ought to possess," thus separating it from sexual lust.[115] Abelard (*c.* 1140) rejected the idea of original sin as guilt, for which he was condemned by the Council of Sens. Thomas Aquinas (*c.* 1270) taught that through his sin Adam lost the ability to keep his inferior impulses in submission to reason without losing his reason, will, and passions. He rejected the role of sexual lust. In the continuing discussion, the Augustinians and Franciscans retained the Augustinian rigorism, while others sided with Abelard in denying the guilt, recognizing only its punitive consequences. However, since Pius V (1856), the Roman Catholic Church has officially held to the Thomist teaching. Luther and Calvin adopted the Augustinian view, equating original sin with sexual lust and claiming that it completely destroys human freedom and persists even after baptism. Although Protestant Liberalism and Roman Catholic Modernism abandoned the doctrine of original sin almost completely, it has been strongly reaffirmed by orthodox Roman Catholic theologians and the Reformed Barthian school.

What I find abhorrent in the doctrine of original sin, whether Augustinian, Thomist, or Barthian, is that it presents us with a fatalistic and pessimistic view of life. It portrays human sinfulness as an inherited disease, as indicated by the German term

Erbsünde. It has found a "scientific" equivalent in the *selfish gene theory* of atheistic biologist Richard Dawkins, who states that "successful genes must be ruthlessly selfish."[116] Although the struggle for survival undoubtedly operates in all living beings, including humans, this cannot in my opinion be seen as the deepest ground of human sinfulness. The biblical arguments for the doctrine of original sin show a double weakness. First, it can be said that, in Romans 5:12-21, Paul wants only to illustrate the superiority of grace over the power of sin, and that understanding these words as support for original sin is taking them out of context.[117] Secondly, proponents of original sin seem to forget that the author of Genesis 3 was reflecting about his experience that all humans are prone to sin, to rebel against God, to grasp for equality with God. In order to explain this universal human attitude, he composed the powerful myth about the first human pair. However, it is not permissible to turn the myth around and to literalize it by claiming that all subsequent human sin was derived, inherited, from Adam.

Another weakness of the doctrine of original sin is that even its modern proponents neglect what we have learned from science about human evolution. We now know that humans developed gradually over a period of 6 million years from *Australopithecus* via *Homo habilis* and *Homo erectus* to *Homo sapiens*, in body as well as mind. Along with this biological evolution there appears to have been a religious and moral evolution.[118] In addition, the theory of evolution tells us that a new species does not originate in a single pair only, but in at least hundreds of individuals. The conclusion is that there cannot have been a single, first human pair such as Adam and Eve in history. This does not deny the truth in the story about Adam and Eve in Genesis 2 and 3, but merely affirms that this story must be seen as a message in mythical (non-historical) form. However, it makes the idea of a hereditary transmission of human sinfulness from Adam to all subsequent generations of humans untenable. The two ideas of *creatio ex nihilo* and original sin have proved to be remarkably tenacious in the history of Christian theology. However, I like to think that chaos theology offers a more satisfactory explanation for the existence of human evil (section 3.5).

Chaos Theology

Predestination

The doctrine of *predestination* was formulated by Augustine in an overreaction to the teaching of the Irish monk Pelagius (*c.* 400). The latter claimed that a person takes the initial and decisive step towards salvation by his own efforts, apart from the assistance of divine grace. In refuting this, Augustine based his position on these words of Paul – "those whom he foreknew he also predestined" (Romans 8:28-30) – and claimed that this means that God decrees the election and non-election of individuals. Calvin made predestination a cornerstone of his system, rejecting the universal saving will of God and maintaining that Christ's atoning death was offered only for the elect. This obviously clashes with the idea of human free will, and so discussion about predestination continues. Reformed theologian Emil Brunner starts his discussion with the words "How terrible and paralysing is all talk of predestination."[119] Anglican theologian John Macquarrie offers a *via media* between Pelagianism and predestination by affirming God's initiative in the work of salvation and the operation of the Holy Spirit in us, but stressing human freedom in accepting this gift.[120] He quotes Paul's words "Work out your own salvation with fear and trembling; for it is God who is at work in you, enabling you both to will and to work for his good pleasure" (Philippians 2:12-13). This seems to me to provide an adequate balance between divine initiative and human free will. In addition, chaos theology combined with the theory of chaos events suggests that God leaves his evolving creation and his creatures a large degree of freedom, intervening only in order to keep creation going towards the goal he has set for it.

3.4 Evil in *Creatio ex Nihilo* Context

Before turning to chaos theology, I shall briefly review the attempts that have been made to find an explanation for the problem of the existence of evil in the creation of a good and powerful God, in *creatio ex nihilo* context. In his book *Evil and the God of Love,* John Hick has reviewed the various explanations that have been advanced over the centuries.[121] He groups them in two models, the Augustinian and the Irenaean model.

There are four key points to the Augustinian model:

1. Evil is absence of good (*privatio boni*). To this I say that defining evil as "absence of the good" may separate evil from God, yet it does hardly justice to the reality of evil.

2. The created world is wholly good, so evil does not come from God. I claim that this is a wrong exegesis of the phrase "God saw that it was good" in Genesis 1; a more correct explanation is that this means that the initial creation satisfies God's ultimate purpose for it (section 1.3).

3. Evil originates from our misuse of freedom. My objection is that this neglects physical evil, and that by introducing the idea of predestination, Augustine still makes God responsible for evil.

4. Physical evil arose because Adam's sin corrupted nature. My objection is that this overlooks the fact that natural disasters and disease preceded the appearance of humans. Dinosaurs already suffered from arthritis!

The Irenaean model is less harsh in maintaining that evil exists ultimately within God's good purpose. God could have created differently, but he knew that early humans were too immature to receive, contain and retain perfection. Original sin and predestination do not figure in this model. My objection is that while this upholds God's goodness, it compromises his omnipotence.

Both models concentrate on moral evil, but neither offers much insight into the problem of physical evil. Both models, particularly the Augustinian model, show little awareness of the evolutionary nature of creation; this surprisingly seems to be true even for most post-Darwinian theologians. An exception is Dutch Calvinist theologian H.M. Kuitert who, in his book *I Have My Doubts*, applies evolution in his discussion of theodicy, but concludes that "evolution is an unpredictable process: it has no purpose."[122] So he does not integrate it in his further discussion of "the riddle of the good creation."[123] Since he still interprets "good" as indicating the quality of the present, unfinished creation, he encounters more problems and rejects more traditional explanations than he offers solutions. His conclusions[124] that "Evil is part of life and we have to put up with it and that God can turn to good what we had thought evil" can hardly be considered to constitute a satisfactory solution of the problem of the existence of evil.

Anton Houtepen, a Dutch Roman Catholic theologian, reviews the positions of the Greek philosophers as well as Augustine, Thomas Aquinas, Luther, Leibnitz, and Kant.[125] In the end he reaches two conclusions: (a) the question resounding in the problem of the existence of evil indicates an awareness in humans of the "possible good"; and (b) the entire human activity of religion, art, science, and technology stems from this quest for the good.[126] These conclusions, right as they may be, do not offer a satisfactory explanation for the theodicy problem, in my opinion. However, I do agree with his vigorous denunciation, based on a study of the story of Job, of the idea that the evil in the world is God's punishment for our sins.[127]

Mark Worthing, whom we already encountered in section 1.2, considers the theodicy problem in the context of contemporary physics.[128] He distinguishes four types of explanation (or rationalization) that have been advanced:

1. *Gubernatio permissa:* God, for reasons unknown, governs the world in such a way that evil is simply permitted to have free rein.

2. *Gubernatio impedita:* God governs the world so as to restrain evil somewhat, but is unable to abolish it.

3. *Gubernatio determinata:* God limits the extent of evil or disorder, so that it cannot exceed a certain determined limit.

4. *Gubernatio directa:* God directs the evil in the world toward a higher or final purpose or goal. This is the Irenaean model, of which Schleiermacher shows himself to be a proponent, when he says, "Sin has been ordained by God; for otherwise redemption itself could not have been ordained."[129] This amounts to causing a shipwreck in order to allow the staging of a rescue operation; I find this theologically unacceptable.

Worthing calls the first two models *dysteleological* (lacking a purpose, and leading only to further disorder), and the last two *teleological* (evil is part of a purposeful process and thus confined to certain established limits). He then connects evil, both physical and moral evil, with the concept of *entropy,* the physical measure of disorder in a natural system. He bases his observations in part on Robert Russell, who notes that entropy and evil are both dependent upon being, and that both lack independent existence.[130] Without order, disorder has no meaning or existence; similarly, without good, evil has no independent existence. I note in passing that disorder may depend on a theoretical concept of order (entropy is a relative quantity), but that a state of perfect order does not exist in any real system in our world. Worthing sees here an analogy with the Irenaean idea that the good depends on the existence of a certain amount of evil in the world, which thus becomes a suitable place for soul-making.

It is interesting to note that some theologians do connect chaos with creation. Thus Philip Hefner writes, "Chaos is the womb of creativity.... Creation and chaos belong together by nature."[131] Rabbi Harold Kushner, author of the bestseller *When Bad Things Happen to Good People,* sees creation as God's ordering of initial chaos, and he recognizes a remaining element of chaos, symbolized in the sea monster Leviathan in Job 41.[132] However, he does not recognize the evolutionary course of creation as God's battle with remaining chaos. So to the question "Why does God not intervene?" Kushner can only reply that God cannot do everything, that God is suffering with us. Brian Hebblethwaite rejects the idea of an initially perfect creation and assumes continuing creation; however, by implicitly retaining *creatio ex nihilo,* he fails to understand the origin of evil, although he has important things to say about "coping" with evil.[133] Edward Schillebeeckx calls *creatio ex nihilo* "clumsy words and images for expressing that God's work transcends our thinking,"[134] and relates evil to the finitude of all that is created, in the sense that the latter provides the possibility for evil to arise. However, in a private discussion he admitted to me that he has no explanation for evil.

In the end I believe that it is fair to say that the problem of evil cannot be explained in the context of *creatio ex nihilo.* As John Sanford says at the end of his book on evil, "The problem of evil is unresolved in Christian theology."[135] In the next section I shall try to present a solution of this problem by means of chaos theology.

3.5 Evil in Chaos Theology

In chaos theology, which is based on the biblical ideas of creation from an initial chaos and the abolition of a remaining element of chaos on the last day, I propose that this remaining chaos element is the source of evil, the physical evil of natural disasters and illness, as well as the moral evil that humans commit against each other and against nature (ecological crisis). Evil is not created but is inherent in the remaining element of chaos. It is a characteristic of the ongoing creation (*creatio continua*), in which remaining chaos is pushed back and ordered until at the last Day it will be definitively abolished (section 1.3). Humans are tempted to submit to chaotic thinking and thus commit morally evil deeds. Yet, knowing the difference between good and evil and having freedom of will, they remain fully responsible for their actions. Physical evil is simply the consequence of the presence of the chaos element in the created world. This explanation of evil seems to me to be more satisfactory than Augustine's *privatio boni*, Barth's *das Nichtige*, Moltmann's "annihilating nothingness, which persists in sin and death," or any of the other explanations outlined in section 3.4. Evil need not be attributed to God anymore or to the effects of the sin of a mythical, proto-human Adam. The created world can be seen and enjoyed as wholly, though not absolutely, good in its God-intended goal.

How can evil come from remaining chaos? I consider chaos itself to be morally neutral. However, both humans and nature are under its influence, and this may lead to moral and physical evil, e.g., "chaotic thinking" may lead humans to evil behaviour. Paul seems to express this in Romans 7:15 ("I do not understand my own actions. For I do not do what I want, but I do the very thing I hate"). In section 3.6 I explain that diseases like cancer and mental illness can be seen as an expression of remaining chaos in the universe. The same can be said for natural disasters such as earthquakes, volcanic eruptions, storms and floods.

This seems to fit in many respects with the thinking of Harold Kushner cited in the previous section.[136] However, to the question "Why does God not intervene?" Kushner can only reply that God cannot do everything, that God is suffering with us. My response, however, is that this is an unsatisfactory view of God's action, that God does intervene, not in instantly curing each of our ills, but in his ongoing battle with the Leviathan of remaining chaos. Where moral evil is concerned, the evil perpetrated by us, God seems to respect us more than we do ourselves: he has made us his image-bearers and given us freedom, so that we are responsible for our sins, and we can and should do something about them ourselves. To me it seems clear that the key elements in an adequate *theodicy* are chaos theology and the evolutionary view of creation (continuing creation). Borrowing the slogan of the Dutch tax service in its television information spots – "We cannot make it more pleasant, but we can make it easier" – we can say that evil does remain just that, but chaos theology makes it more understandable. Such understanding is not only of theological importance, but is also of great pastoral significance, as I will explain in the next section.

3.6 A Theology of Disease

Recent polls suggest that above family life, work, money, and religion, health is the number one concern of persons in the Western nations. Even though as a Christian I disagree with this ranking, it shows how essential it is to address the topic of health and disease in any attempt to reconcile the scientific and theological world views. Remarkably, neither dogmatic theologians nor scientist-theologians have given much attention to this topic, which is of such importance in human life.[137] There is a great need for a "theology of illness" in view of the persisting tendency to ascribe disease to sin or to divine punishment for sin, leading to misplaced guilt feelings in many seriously ill people. Even such an eminent Christian as Dame Cicely Saunders, the founder of the hospice movement, held this conviction.[138]

Biblical teaching

In the Old Testament, health was seen as a divine gift, and disease was regarded as divine punishment for sin or disobedience (e.g., Exodus 4:11; Deuteronomy 32:39). This idea persisted in the popular mind throughout the entire biblical period.

However, Jesus firmly rejects the idea that disease is God's punishment, either for personal sins or those of the parents (John 9:3). He sees the individual as an essential unity of body and mind, and sees disease as the result of evil producing an imbalance in the body-mind unity, as a manifestation of evil operating in the recesses of the mind. In his view, disease violates the divinely established order. Hence, in his healing acts Jesus pays close attention to the mind of the sick, often linking healing with forgiveness of sins (Mark 2:2-11). In his encounter with the Samaritan woman (John 4:7-26), a casual conversation turns into a powerful therapeutic analysis of her emotional conflicts, a superb example of non-directive counselling. Jesus thus anticipates current holistic medicine and psychotherapy. His healing acts are a spontaneous expression of his compassion as well as a sign of the kingdom of God. In the early Church his healing ministry is continued with the laying on of hands (Acts 9:12), with prayer and with anointing (James 5:14-15).

Modern understanding and treatment of disease

I shall use cancer as a model because of its prevalence and high mortality rate and because much is known about its molecular mechanism.[139] The first step in the cancer process is the transformation of a normal cell into a cancer cell. Normal cells in our body reproduce only when instructed to do so by neighbour cells. Two genes, the proto-oncogene and the tumour suppressor gene, control the cell cycle, the process of growth and division of a cell. Mutation of the proto-oncogene turns it into an oncogene, which causes the neighbour cell to multiply excessively. Mutation of the tumour suppressor gene cancels its growth-inhibiting effect on the neighbour cell. These two effects cause uncontrolled, rapid division of the neighbour cell, transforming it into a cancer cell. Cells have two defense systems against such runaway growth: (a) *apoptosis* or cell suicide, which is essential for removing unwanted cells in embryonic development, but is inactive in normal adult cells; and (b) *telomere shortening* (a telomere is a DNA segment at the end of a chromosome that is shortened every time the chromosome is replicated, limiting normal cells to 50 to 60 divisions). "Successful" cancer cells are able to inactivate the protein that triggers apoptosis, and to activate the enzyme *telomerase* that rebuilds shortened telomeres. Having thus outwitted these two defense systems, the cancer cell multiplies and forms a primary tumour. When this reaches a diameter of about 1.6 mm, the tumour begins to suffer from oxygen deficiency. This stimulates *angiogenesis*, the formation of small blood vessels, and allows the tumour to continue its growth.

The second step is the dislodging of one or more tumour cells, leading to *metastasis*. In a normal tissue the cells produce certain proteins that make them adhere to neighbor cells and, in the case of an epithelial tissue, to a membrane of a blood vessel or lymph vessel. Cancer cells lack the adhesion proteins (or their receptors), and thus can move out of the tissue in which they were formed. At the same time, the enzyme metalloproteinase in the tumour cell is activated and excreted. This enzyme drills a hole in the membrane through which the tumour cell can enter the vessel. The cell is then carried with the blood or lymph flow until it reaches a capillary bed, where it is stopped, adheres, and forms a secondary tumour. This process can be repeated with other tumour cells and thus lead to widespread metastasis. When primary or secondary tumours disrupt essential body functions, the patient dies.

There are three types of treatment to remove or kill the tumour: surgical removal (if there is no metastasis), radiation and chemotherapy. These therapies, singly or in combination, can in many cases greatly increase life expectancy with an acceptable quality of life. However, the problem is that total removal or destruction of all cancer cells in a patient is very difficult to achieve, if not impossible, so a true cure is still rare.

Theological interpretation

Disease may be considered as a form of physical evil. After all, disease occurs in plants and animals and thus appears to have arisen long before the appearance of humans. As a matter of fact, the evolutionary process appears to play an important role in the development of disease.[140] From our current understanding of the molecular mechanism of cancer it can be concluded that the primary event is the random mutation of a single gene in one body cell, which we can consider to be a chaos event (section 1.5). This event has several and important consequences as we have seen: unlimited cell division, dislodging of the transformed cell, formation of blood vessels in the early tumour permitting its further growth, secondary tumour formation in metastasis, and, finally, death of the patient when growing tumours disrupt vital organ functions.

The cancer process is the derailment of a very complex, orderly, coordinated functioning of many genes, enzymes, hormones, messenger proteins and receptors existing in our body cells under normal conditions. This order has been established by the Creator in the course of evolutionary creation and is established anew in each individual, owing to the genetic system present in its cells. A chaos event, the random mutation of one gene in one cell, can make this order degenerate to chaos on the cellular level. The same can be said for all diseases in which a normal physiological mechanism is derailed. It may also apply to psychiatric disease, various types of which are ascribed by psychotherapist-theologian Eugen Drewermann to the fear of being thrown back into primordial chaos, of which he sees a remaining element in our world.[141]

In terms of chaos theology we may conclude that cancer and other diseases are caused by the remaining chaos element disturbing the order established by the Creator in the evolutionary creation process. This theological interpretation is in accord with the message of Jesus that disease is a manifestation of evil, a disturbance of divine order, but is not a punishment for sin of the sick person or his parents (John 9:3) or to make us grow spiritually through suffering (although spiritual growth may be a salutary consequence of suffering). Guilt (but not divine punishment) can only be spoken of when the disease is due to our negligence, e.g., cirrhosis of the liver through alcohol abuse, or HIV infection through unprotected sex with multiple partners or through the re-use of hypodermic syringes. Insofar as chaos events play a role in disease, we may assume that divine and human spirit can influence both the beginning and the course of an illness. This legitimates prayer and the use of the sacrament of healing, and

3: The Problem of Evil

may account for medically unexplained cures and remissions. This will be discussed below.

Curing or healing

It is increasingly recognized that current cancer treatment, which is almost exclusively directed against the tumour (curing), neglects the role that our mind may play (healing). The mechanism for the interaction between mind and body is beginning to be understood scientifically.[142] From the brain cortex, the seat of the mind, nerves run to the hypothalamus, which, upon stimulation, secretes activating substances to the nearby pituitary gland, making it secrete hormones that affect several body systems, including the immune system. When the mind is in a bad state due to stress, conflict, or guilt feelings, the immune function may be impaired. The transformation of a normal body cell into a cancer cell through random mutation appears to occur fairly frequently. Normally our immune system will recognize such a cell as abnormal and destroy it, but this may not happen when the immune system is impaired. This theory is supported by the observation that renal transplant patients, who must be kept on immunosuppressive drugs to prevent rejection, have a highly increased rate of cancer incidence (10% increase after one year, 50% after 10 years).[143]

These observations suggest that more attention needs to be given to "healing." There are now institutes offering programs to stimulate the self-healing capacity of patients so as to supplement (but not replace) conventional medical treatment.[144] The patients are helped to liberate themselves from wrong ideas about guilt, sin, and punishment, and to express their feelings of anger and anxiety, allowing them to reintegrate body and mind. In some studies an enhancement of the immune function has been observed. Although these institutes operate on a non-religious basis, their approach resembles Jesus' practice of healing coupled with the offer of forgiveness. Some recent polls indicate that in our secularized Western world a relation between faith and healing is still widely recognized.[145] In a 1996 *Time*/CNN poll, 82% of Americans professed a belief in the healing power of personal prayer. In a Yankelovich poll, 99% of the 269 physicians present at the annual meeting of the American Academy of Family Physicians were convinced that religious belief can contribute to healing. A poll by *Nature* in 1997 showed that 40% of biologists, physicists and mathematicians believe in a God who answers prayers. Statistical studies show the importance of religious commitment (measured as attendance and participation) in combatting cancer and other diseases.[146]

The practice in the early Church of the laying on of hands with prayer for healing has been revived during the 20th century in Anglican churches. Even when this does not lead to a cure, it may provide healing in the sense of promoting peace of mind and the assurance that God will guide us through the final stage of earthly life towards eternal life, the ultimate life for which we are created and destined. In my opinion the sacrament of healing deserves a place in the Sunday Eucharist after the distribution of the consecrated elements, in the midst of the congregation. A form which shows a proper balance between curing and healing is provided in the 1979 *American Prayer Book*.[147]

3.7 Future and Destiny

Scientific view of the future

As a fitting conclusion to this monograph I consider what chaos theology can say about eschatology. First I wish to look at what science can tell us about the future of the universe. If we assume a "flat" universe, as now seems most likely (see section 2.2), then this means that in billions of years from now the universe will stop expanding, all stars will burn out and turn into black holes, which will in the end disintegrate into photons, neutrinos, electrons and positrons – a cold, dark, lifeless universe.[148] Long before then, some 5 billion years from now, our Sun will turn into a red giant, raising the temperature on Earth to about 1300°C, thus ending all life on Earth. Before this happens, a large asteroid may hit the Earth, as occurred twice before, and destroy life by the direct impact or by subsequent cooling from dust clouds circling the Earth.[149] And now we are aware of the even more immediate threat of global warming, which may decimate humankind. Science offers a bleak picture of the future of the universe and its creatures. From the order that has been built up in some 15 billion years from the chaos of the big bang, the universe seems doomed to return to complete, lifeless chaos. This would mean that the entire cosmic and biological evolutionary process was utterly futile.

Biblical view of the future

In sharp contrast to the somber view that science offers, the biblical view of the future is joyful. From the biblical perspective, future becomes destiny, the end of time as we know it, the last Day. This subject in theology is called eschatology, the teaching about the *eschata* or last things. Isaiah speaks of God as the first and the last (Isaiah 44:6; Isaiah 48:12) and he brings the message of a new creation (Isaiah 41:17-20; 43:18-21): "Behold, I am doing a new thing; now it springs forth, do you not perceive it?" (Isaiah 43:18-19). Isaiah also speaks about the new heaven and the new earth that God will make (Isaiah 66:22). The Old Testament prophets forecast the coming of a Messiah, who will bring healing to broken humanity and to the entire universe (e.g., Isaiah 11:1-9; Jeremiah chapters 5 and 6).

The New Testament authors develop the idea of a new creation by integrating it with their understanding of the central role of Christ in salvation. The title applied by Isaiah to God is given to Christ: "I am the Alpha and the Omega, the first and the last, the beginning and the end" (Revelation 22:13). Christ stands at the beginning as the creative Logos, at the end as the inaugurator of the new creation: Alpha and Omega, the first and last letters of the Greek alphabet. The central New Testament message then is that in Christ God has already introduced the new creation, which the Old Testament prophets were expecting. Yet, the new creation is also a promise for the end of time when there will be a new heaven and a new earth and no more evil and death: "Then I saw a new heaven and a new earth; for the first heaven and the first earth had passed away, and the sea was no more"

Chaos Theology

(Revelation 21:1). The sea that is abolished symbolizes the remaining element of the initial chaos in terms of chaos theology. The paradox of a new creation, already introduced and yet to come in future, can only be resolved by seeing Christ's death and resurrection as the decisive victory, which in God's good time will bring the final liberation of his creation. For this paradox Charles Dodd introduced the term *realized eschatology*.[150]

So the New Testament makes a threefold claim: (a) the promised Messiah is Jesus of Nazareth, the Christ who is the creative Word of God operating in creation; (b) Jesus died for us humans and this world and rose as the firstborn of the dead, assuring the world and us of reconciliation; and (c) Christ will return on the last Day at the transformation of the created universe into the kingdom of God. His resurrection changes <u>e</u>volution into <u>r</u>evolution, and, in the new kingdom, death will change into eternal life, remaining chaos into perfect order. We are now living between the decisive event of the resurrection and the final event of the coming kingdom. Through Christ this world is transformed into the new kingdom, in which humans may live in a new relation to God and therefore to their fellow humans and to the entire creation. Looking backward, the Christian community can say, "In Christ all things were created," and looking forward, it may joyfully exclaim, "God will sum up all things in Christ, and will make all things new."

To use chaos theology terminology, we may assert that at that time the Creator will banish the remaining chaos element, thus abolishing both moral and physical evil, and he will fulfill his original intention and perfect his creation; future will become destiny. While it is not given to us to know the time, I think we can safely assume that God will bring this about before degeneration or catastrophe can destroy the universe and all life in it.

Why these different views of the future?

The simple answer to that question is that science can only consider this world and does not know of any energy source outside it. Thus, according to the second law of thermodynamics, this universe must wind down with the arrow of time and return to disorder, utter chaos. Science cannot speak about the value or futility of the evolving cosmos.

The biblical view assumes that the Creator will not abandon his creation and will continue to "energize" it. Thus, the cosmic entropy will continue to go down until a state of full order is reached in which the remaining element of chaos has been abolished. Our assurance that this will happen is that it is unthinkable that God would create this universe to let it disintegrate. For us, as the only beings created in the likeness of God, there is the challenge to believe this and thus to become part of the perfected creation.

What can we say about judgment on the last day? Judgment is an ominous word to many who recall the lurid medieval pictures of hell. However, the image of a vengeful and retributive God is foreign to the New Testament (except 2 Thessalonians 1:5-10 and Revelation); God's justice is loving, creative, restorative, reconciling, and healing. About one third of those who have had a near-death experience report seeing their entire life as in a 3-D movie with an experience of judgment.[151] Therefore, it seems to me that at death we are not judged so much as that we judge ourselves in answer to the question Am I able and desirous to live in God's presence or not? In the intermediate period between our death and the last day there will be an opportunity for further spiritual growth, until at the last Day we will judge ourselves definitively to either living eternally in God's presence (heaven) or to existing forever in his absence (hell).[152]

3.8 Summary and Conclusions (Chapter 3)

The problem of evil looms large in the mind of believers as well as non-believers today. The occurrence of evil in the "good" creation of an omnipotent and just God has remained an unsolved problem, because the doctrine of creation out of nothing (*creatio ex nihilo*) makes God responsible for evil. Introducing Satan does not help. If he is under God's control (Job), then God remains responsible; if he is not under God's control, we are back to the demiurge of gnostic dualism. How can we explain the existence

Chaos Theology

of evil in God's creation without lapsing into dualism or compromising either God's goodness or his omnipotence?

Humans are ambivalent beings, living as God's image bearers and co-creators, but also grasping for equality with God. The latter activity is the root of every kind of moral evil. This insight describes the occurrence of moral evil, but it does not explain why humans behave in this way, or how God can allow this to occur. The doctrine of original sin, conceived by Irenaeus, suggests that evil came into the world through the sin of Adam and was transmitted to all further generations of humans as a kind of hereditary disease. Augustine coupled this with sexual lust and added to it the doctrine of predestination, basing his thinking on Romans 8:28-30. This clashes with the idea of human free will, which is described in Philippians 2:12-13.

Chaos theology, combined with the theory of chaos events, suggests that God leaves his evolving creation a large degree of freedom, intervening only in order to keep it going towards the goal he has set. Against the doctrine of original sin I claim that the biblical arguments show a double weakness: Romans 5:12-21 is taken out of context, and the mythical story in Genesis 3 should not be turned around to claim that all subsequent human sin was inherited from Adam. Further, I note that our insight in biological evolution indicates that there can never have been a single first human pair, but that there was a gradual physical, religious and moral development during human evolution.

A review of solutions offered for a theodicy in the context of *creatio ex nihilo* shows that none can be considered satisfactory. They can be grouped into two basic models. The Augustinian model (seeing evil as the absence of good, as "nothing") may separate evil from God, but with the doctrine of predestination it still makes God ultimately responsible for evil. Explaining physical evil by assuming that Adam's sin corrupted nature ignores the fact that this form of evil existed long before the appearance of humans. Considering life's trials as punishment for Adam's sin contradicts God's justice and our individual responsibility. The Irenaean model avoids original sin and predestination, but it reduces God's omnipotence by maintaining that evil exists ultimately within God's good purpose. Both models neglect the evolutionary nature of creation. Considerations by modern theologians do not offer a satisfying theodicy either.

Chaos theology and the evolutionary view of creation are the key elements in my theodicy. Creation is seen as an evolutionary process of ordering from initial chaos. The remaining element of chaos is expressed in physical and moral evil. Evil is thus neither created by God, nor the result of Adam's sin, but is inherent in the remaining element of chaos. The created world is good in the sense that it satisfies God's ultimate purpose for the ongoing creation. To the question "Why does God not intervene?" I reply that God does intervene, not in instantly curing all our ills, but by "healing" his creation and his creatures in his ongoing battle with remaining chaos.

Disease as a form of physical evil has not received much attention from theologians, although the Bible has much to say about it. Jesus firmly rejects the Old Testament idea of disease as divine punishment for sin. He sees the individual as a unity of body and mind, and disease as the result of an imbalance between them, so he ministers both to mind and body in his healing acts. Science shows that cancer results from a chaos event – the random mutation of a single gene in a single body cell – causing finely-tuned order on the cellular level to revert to chaos. Expressed in terms of chaos theology, the remaining element of chaos is acting against the order established in the evolutionary creation process. Current treatment of cancer, aimed at "curing" by removal or destruction of the tumour, neglects the role of the mind in disease. "Curing" is only a part of the "healing" of the body–mind unity. The healing acts of Jesus, aimed at healing of the body-mind unity, are continued in the laying on of hands with prayer. This sacrament can thus be meaningfully used in supplementing conventional medical treatment. Even when a cure does not result, it may provide healing in the sense of promoting peace of mind, reconciliation, and the assurance that God will guide us through the final stage of earthly life towards eternal life, the ultimate life to which we were created and destined.

Chaos Theology

I concluded with a consideration of eschatology, the future and destiny of the universe and ourselves. Science gives us the somber picture of a degenerating universe with all life coming to an end through the burning out of all stars, including our Sun, or possibly sooner on Earth by the impact of a large asteroid or by the effects of global warming. In contrast to this, the biblical view is joyful in speaking about a new heaven and a new earth, the transformation of the present world into the kingdom of God, the perfection of God's creation. Christ stands at the beginning as the creative Logos, at the end as the inaugurator of the new creation: Alpha and Omega. In the terminology of chaos theology, the remaining element of chaos (the sea) will be abolished and with it all evil and death. The contrast between the two views is due to the fact that science can only consider this world and does not know of any energy source outside it, so the universe must degenerate to chaos. The biblical view assumes that God keeps energizing his universe and thus driving entropy to zero. Judgment is seen as self-judgment on the basis of the reports of people who have had a near-death experience. On the last Day we will judge ourselves by answering the question: Am I able and desirous to live in God's presence or not?

In this monograph I have argued for the replacement of the traditional doctrine of *creatio ex nihilo* (creation from nothing) by chaos theology (creation from initial chaos with a remaining element of chaos). Chaos theology, in combination with the physical theory of chaos events, can greatly contribute to the reconciliation of the scientific and theological world views. It can cast fresh light on other crucial aspects of theology, God's activity in the world, Christ's incarnation and saving action, the problem of evil, the theological understanding of disease, and eschatology, while doing away with the awkward doctrines of original sin and predestination.

Notes

[1] Sjoerd L. Bonting, *Schepping en Evolutie: Poging tot Synthese* (Creation and evolution: Attempt at synthesis) (Kampen, NL: Kok, 1996, 1997). English text available upon request.

[2] Sjoerd L. Bonting, "Chaos Theology: A New Approach to the Science-Theology Dialogue," *Zygon Journal of Religion & Science* 34, no. 2 (1999): 323–332.

[3] Gerhard May, *Creatio ex Nihilo; The Doctrine of "Creation out of Nothing" in Early Christian Thought* (Edinburgh: T&T Clark, 1994), 26–38.

[4] Ellen van Wolde, *Stories of the Beginning* (London: SCM Press, 1996).

[5] Gerhard May, *Creatio ex Nihilo,* 39–61.

[6] Ibid.

[7] Ibid.

[8] John Polkinghorne, *Science and Christian Belief* (London: SPCK, 1994), 76.

[9] Claus Westermann, *Genesis 1–11. A Commentary*, trans. John J. Scullion, S.J. (London: SPCK, 1994), 110, 121.

[10] David A.S. Fergusson, *The Cosmos and the Creator; An Introduction to the Theology of Creation* (London: SPCK, 1998), 23.

[11] Mark W. Worthing, *God, Creation, and Contemporary Physics*, (Minneapolis: Fortress Press, 1996), 105, 110.

[12] John Polkinghorne, *Science and Creation* (London: SPCK, 1988), 59.

[13] Arthur Peacocke, *The Idreos Lectures: The Quest for Christian Credibility* (Oxford: Harris Manchester College, 1997), 31.

[14] Mark W. Worthing, *God, Creation, and Contemporary Physics,* 79–110.

[15] Karl Barth, *Church Dogmatics* III, 2, ed. G. Bromiley and T.F. Torrance (Edinburgh: T&T Clark, 1960), 154ff.

[16] Emil Brunner, *The Christian Doctrine of Creation and Redemption*, Vol.II, *Dogmatics.* 4th ed., (London; Lutterworth, 1960), 9–21.

[17] Paul Tillich, *Systematic Theology*, vol. 1 (Chicago: University of Chicago Press, 1951), 188.

Chaos Theology

[18] Mark W. Worthing, *God, Creation, and Contemporary Physics*, 75.

[19] Ibid., p. 75.

[20] Keith Ward, *God, Faith and the New Millennium: Christian Belief in an Age of Science* (Oxford: Oneworld, 1998), 53–59.

[21] Jürgen Moltmann, *God in Creation* (San Francisco: Harper, 1991), 86–93.

[22] Ibid., 90.

[23] David A.S. Fergusson, *Cosmos and Creator*, 27.

[24] These include John Hick, *Evil and the God of Love* (Glasgow: Collins, 1979); Harold S. Kushner, *When Bad Things Happen to Good People* (London: Pan Books, 1982); John A. Sanford, *Evil: The Shadow Side of Reality* (New York: Crossroad, 1989); David A.S. Fergusson, *Cosmos and Creator*, 77–87; Hermann Häring, *Das Böse in der Welt; Gottes Macht oder Ohnmacht?* (Darmstadt: Primus Verlag, 1999); Brian Hebblethwaite, *Evil, Suffering and Religion* (London: SPCK, 2000).

[25] Brian Hebblethwaite, ibid., 110.

[26] Sjoerd L. Bonting, *Schepping en Evolutie*, 34–36; Ellen van Wolde, *Stories*, 188–194.

[27] This is the NRSV version; the older version, as in the RSV (*In the beginning God created the heavens and the earth. The earth was without form and void...*), though grammatically possible since Hebrew does not have the subordinate clause, makes no theological sense, as it would have God create chaos.

[28] Ellen van Wolde, *Stories of the Beginning*.

[29] Charles H. Long, *Primitive Religion*. Academic American Encyclopedia (electronic version 11.0 RL). Grolier Inc., Danbury, CT, 1996.

[30] Claus Westermann, *Creation*, trans. John J. Scullion, S.J. (London: SPCK, 1974), 4–15.

[31] Philip Hefner, "The Evolution of the Created Co-Creator," in *Cosmos as Creation: Theology and Science in Consonance*, ed. Ted Peters (Nashville: Abingdon Press, 1989), 226.

[32] G. von Rad, *Genesis: A Commentary*, trans. J.H. Marks (London: SCM Press, 1951), 49.

[33] Karl Barth, *Church Dogmatics*, III, 154 ff.

Notes

³⁴ Martin Kiddle, "The Revelation of St. John," in *Moffatt New Testament Commentary* (London: Hodder and Stoughton, 1940), 411: "The sea personifies the very principle of disorder in the creation.... When the sea vanishes we know that all the imperfections of the first creation have gone with it."

³⁵ John B. Cobb and David R. Griffin, *Process Theology: An Introductory Exposition* (Philadelphia: Westminster Press, 1976), 65.

³⁶ Th.C. Vriezen, *An Outline of Old Testament Theology* (Newton, MA: Branford, 1958), 187.

³⁷ Margaret Wertheim, "God of the Quantum Vacuum," *New Scientist*, 4 October 1997, 28–31.

³⁸ Arthur Peacock, "Chance and Law in Irreversible Thermodynamics, Theoretical Biology, and Theology," in *Chaos and Complexity: Scientific Perspectives on Divine Action,* eds. Robert J. Russell, Nancey Murphy, and Arthur R. Peacocke (Vatican City State: Vatican Observatory Publications, and Berkeley, CA: Center for Theology and Natural Sciences, 1995), 123–143.

³⁹ Tom Stonier, *Information and the Internal Structure of the Universe* (London: Springer Verlag, 1990), 38–41, 70–72.

⁴⁰ Ann Gibbons, "When It Comes to Evolution, Humans Are in the Slow Class," *Science* 267(1990): 1907–1908.

⁴¹ Pierre Teilhard de Chardin, *The Phenomenon of Man* (London: Collins, 1959), 257–264, 268–272, 288–289.

⁴² Paul Tillich, *The Courage to Be* (Douglas, U.K.: Fontana, 1962), 41–68.

⁴³ Eugen Drewermann, *Exegese en Dieptepsychologie* (Exegesis and depth psychology) (Zoetermeer, NL: Meinema, 1993), 27.

⁴⁴ Stephen J. Gould, "The Evolution of Life on the Earth," *American Scientist* 271, no. 4 (April 1994): 62–69. Similar views are presented by Michael Ruse, *Monad to Man; The Concept of Progress in Evolutionary Biology* (Cambridge, MA: Harvard University Press, 1996).

⁴⁵ Jacques Monod, *Chance and Necessity* (London: Collins, 1972).

⁴⁶ Robert J. Russell, Nancey Murphy, and Arthur R. Peacocke, eds., *Chaos and Complexity: Scientific Perspectives on Divine Action* (Vatican City State: Vatican Observatory Publications, and Berkeley, CA: Center for Theology and Natural Sciences, 1995).

⁴⁷ Wesley J. Wildman and Robert J. Russell, Chaos: A Mathematical Introduction with Philosophical Reflections, *Chaos and Complexity*, 49–90.

[48] Arthur Peacocke, *Theology for a Scientific Age* (London: SCM Press, 1993), 50–53.

[49] Kathleen T. Alligood, Tim D. Sauer, and James A. Yorke, *Chaos: An Introduction to Dynamical Systems* (New York: Springer-Verlag, 1996); reviewed by J.A. Rial, *American Scientist* 85, no. 5 (1997): 487–488.

[50] John Polkinghorne, The Metaphysics of Divine Action, *Chaos and Complexity*, 147–156.

[51] Arthur Peacocke, God's Interaction with the World, *Chaos and Complexity*, 263–287.

[52] Willem B. Drees, Gaps for God?, *Chaos and Complexity*, 223–237.

[53] John Polkinghorne, *One World: The Interaction of Science and Theology* (1986; reprint, London: SPCK, 1993), 6–25.

[54] Harold K. Schilling, *The New Consciousness in Science and Religion*. (London: SCM Press, 1973), 75, 116–119.

[55] Robert J. Russell, "Is Nature Creation?", in *The Concept of Nature in Science and Theology*, ed. N.H. Gregersen *et al*, vol. 3 of *Studies in Science & Theology* (Geneva: Labor et Fides, 1995), 117.

[56] John Gribbin and Martin Rees, *The Stuff of the Universe: Dark Matter, Mankind and Anthropic Cosmology* (London: Penguin Books, 1995). See also: Donald Goldsmith, "Supernovae Offer First Glimpse of Universe's Fate," *Science* 276(1997):37–38.

[57] J.P. Moreland, ed., *The Creation Hypothesis: Scientific Evidence for an Intelligent Designer* (Downers Grove, IL: InterVarsity Press, 1994), 141–172, Table 4.4.

[58] Paul Davies, *God and the New Physics* (London: Dent, 1983), 179. For direct evidence for a flat universe from background radiation measurements, see George Musser, *Scientific American* 279, no. 3 (1998):13.

[59] Alan H. Guth, *The Inflationary Universe* (Reading, MA: Addison-Wesley, 1997).

[60] Charles Seife, "Echoes of the Big Bang Put Theories in Tune," *Science* 292 (2001):823. The picture has become more complicated by observations of a very early supernova, suggesting an accelerating expansion with notions of antigravity and dark energy ("Revolution in Cosmology, Special Report," *Scientific American* 280, no. 1 (January 1999):27–49; George Musser, "Boom or Bust?", *Scientific American* 281, no. 4 (October 1999):18–19; Charles Seife, "Peering Backward to the Cosmos' Fiery Birth," *Science* 292(2001):2236–2238). However, these conclusions have been faulted by Kenji Tomita in his *local void model* (K. Tomita, *Progress in Theoretical Physics* 105, no. 3 [2001]).

Notes

[61] John D. Barrow and Frank J. Tipler, *The Anthropic Cosmological Principle and the Structure of the Physical World* (New York: Oxford University Press, 1986).

[62] John Polkinghorne, *Beyond Science* (Cambridge: Cambridge University Press, 1996), 87–89. Less polite was Martin Gardner in his review of Barrow and Tipler's book in *The New York Review of Books* (8 May 1986, pp. 22-25): "In my not so humble opinion I think the last principle is best called CRAP, the Completely Ridiculous Anthropic Principle."

[63] J. Richard Gott, "Creation of Open Universes from de Sitter Space," *Nature* 295(1982): 306.

[64] Lee Smolin, *The Life of the Cosmos* (New York: Oxford University Press, 1997).

[65] Joseph Silk, "Holistic Cosmology," *Science* 277 (1997):644.

[66] Charles Seife, "Big Bang's New Rival Debuts with a Splash," *Science* 292 (2001):189–191.

[67] Sjoerd L. Bonting, *Mens, Chaos, Verzoening* [Humanity, chaos, reconciliation] (Kampen: Kok, 1998), 36–37. English texte available on request.

[68] J.P. Moreland, *The Creation Hypothesis,* 164–172, Table 4.5.

[69] Paul Davies, *The Cosmic Blueprint* (London: Unwin, 1989), 131–132.

[70] Virginia Morell, "On the Many Origins of Species," *Science* 273(1996):1496–1502. New species do not only arise through physical separation by mountain ridge, ocean, or river (allopatry), but also side by side after selection of a new food source together with preference for a mate also using this food (sympatry).

[71] Glenn Goodfriend and Stephen J. Gould, *Science* 274, 13 Dec. 1996.

[72] Christine Mlot, "Microbes Hint at a Mechanism Behind Punctuated Evolution," *Science* 272(1996):1741.

[73] David N. Reznick et al, "Evaluation of the Rate of Evolution in Natural Populations of Guppies (Poecilia reticulata)," *Science* 275(1997):1934–1937.

[74] Mary and John Gribbin, *Being Human* (London: Orion Books, 1995), 248–276.

[75] Michael Behe, *Darwin's Black Box: The Biochemical Challenge to Evolution* (New York: Free Press, 1996). He argues from the complexity of various biological systems (vision, immune system, blood clotting) for the operation of "Intelligent Design."

[76] Elizabeth Pennisi and Wade Roush, "Developing a New View of Evolution," *Science* 277(1997):34–37.

[77] Charles S. Zuker, "On the Evolution of Eyes," *Science* 265(1994):742–743.

[78] Wade Roush, "'Master Control'" Gene for Fly Eyes Shares Its Power," *Science* 275(1997):618–619. Two additional eye regulator genes, ey and dac, have been found in Drosophila, which suggests the existence of a hierarchy of such genes.

[79] Richard A. Kerr, "Biggest Extinction Looks Catastrophic," *Science* 280(1998):1007.

[80] Douglas E. Erwin, "The Mother of Mass Extinctions," *Scientific American* 275, no. 1 (July 1996):56–62; A.H. Knoll et al, "Comparative Earth History and Late Permian Mass Extinction," *Science* 273(1996):452–457; Yukio Isozaki, "Permo-Triassic Boundary Superanoxia and Stratified Superocean: Records from Lost Deep Sea," *Science* 276(1997):235–238.

[81] Walter Alvarez, *T. Rex and the Crater of Doom* (Princeton, NJ: Princeton University Press, 1997).

[82] Richard Dawkins, *The Blind Watchmaker* (London: Norton, 1986).

[83] Yves Coppens, "East Side Story: The Origin of Humankind," *Scientific American* 270, no. 5 (May 1994):62–69. A correlation between periods of a cold, dry climate in East Africa 2.8, 1.7, and 1.0 million years ago with major steps in the evolution of African hominids is suggested by Peter B. deMenocal, "Plio-Pleistocene African Climate," *Science* 270 (1995):53–58.

[84] Ann Gibbons, "The Mystery of Humanity's Missing Mutations," *Science* 267 (1995):35–36.

[85] Ann Gibbons, "When It Comes to Evolution, Humans Are in the Slow Class," *Science* 267 (1995):1907-1908.

[86] Robert J. Russell, et al, *Chaos and Complexity*.

[87] Thomas F. Tracy, Particular Providence and the God of the Gaps, *Chaos and Complexity*, 289–324

[88] Nancey Murphy, Divine Action in the Natural Order, *Chaos and Complexity*, 325–357.

[89] George F.R. Ellis, Ordinary and Extraordinary Divine Action: The Nexus of Interaction, *Chaos and Complexity*, 359–395.

[90] John Polkinghorne, The Metaphysics of Divine Action, *Chaos and Complexity*, 147–156.

Notes

[91] Arthur Peacocke, God's Interaction with the World, *Chaos and Complexity*, 263–287.

[92] Hugh Montefiore, *The Probability of God* (London: SCM Press, 1985), 154–165.

[93] Aubrey Moore, "The Christian Doctrine of God," in *Lux Mundi*, ed. Charles Gore (London: Murray, 1890), 99.

[94] Arthur Peacocke, God's Interaction with the World, *Chaos and Complexity*, 263–287.

[95] John Polkinghorne, *Science and Providence: God's Interaction with the World* (1989; reprint, London: SPCK, 1994), 31.

[96] John B. Cobb and David R. Griffin, *Process Theology*.

[97] Arthur Peacocke, *Theology for a Scientific Age*, pp. 157–160.

[98] Hugh Montefiore, *The Probability of God*.

[99] Austin Farrer, *God Is Not Dead* (New York: Morehouse-Barlow, 1966), particularly p. 85.

[100] John B. Cobb and David R. Griffin, *Process Theology*.

[101] Ibid., 63–68.

[102] John Polkinghorne, *Science and Providence*, p. 14.

[103] Keith Ward, *Rational Theology and the Creativity of God* (London: Blackwell, 1982), 229.

[104] Arthur Peacocke, *Theology for a Scientific Age*, 275–279.

[105] John Macquarrie, *Jesus Christ in Modern Thought* (London: SCM Press, 1990), 393. It is interesting that Macquarrie has changed his mind since 1966 (*Principles of Christian Theology*, 258–260), when he did not recognize this problem.

[106] John A.T. Robinson, *Honest to God* (London: SCM Press, 1963), 71.

[107] J.B. Phillips, *The New Testament in Modern English* (London: G. Bles, 1960), 181.

[108] C.F.D. Moule, *The Origin of Christology* (Cambridge: Cambridge University Press, 1978), 47–96.

[109] C.J. den Heyer, *Verzoening, bijbelse notities bij een omstreden thema* (Reconciliation: Biblical notations on an embattled theme) (Kampen, NL: Kok, 1997).

110 Marjoleine de Vos, "Het Eindeloze Zoeken" [The endless search], *NRC Handelsblad*, 22 September 1997.

111 Anton Houtepen, *God, een Open Vraag* (God: An open question) (Zoetermeer, NL: Meinema, 1997), 97.

112 Harold S. Kushner, *When Bad Things Happen to Good People* (London: Pan Books, 1982).

113 John A. Sanford, *Evil: The Shadow Side of Reality*, p. 129: "The problem of evil is unresolved in Christian theology."

114 John Paul II, Encyclical *Fides et Ratio* [Faith and reason], 1999, no. 76.

115 Hermann Häring, *Das Böse in der Welt*, 103.

116 Richard Dawkins, *The Selfish Gene* (Oxford: Oxford University Press, 1976).

117 Emil Brunner, *The Christian Doctrine of Creation and Redemption*, 98–100. He rejects the Augustinian idea of "original sin" as completely foreign to the thought of the Bible (103–107).

118 Sjoerd L. Bonting, *Mens, Chaos, Verzoening*, 154–165.

119 Emil Brunner, *The Christian Doctrine of God*, vol. 1, *Dogmatics* (London: Lutterworth, 1960), 303–339.

120 John Macquarrie, *Principles of Christian Theology* (London: SCM Press, 1966), 297–298.

121 John Hick, *Evil and the God of Love* (Glasgow: Collins, 1979).

122 H.M. Kuitert, *Het Algemeen Betwijfeld Christelijk Geloof* (Baarn, NL: Ten Have, 1992). *I Have My Doubts: How to Become a Christian Without Being a Fundamentalist* (London: SCM Press, 1993), 76.

123 Ibid., 73–75

124 Ibid., 108.

125 Anton Houtepen, *God, een Open Vraag*, 79, 97–103.

126 Ibid., 104–125.

127 Ibid., 91–93.

128 Mark W. Worthing, *God, Creation, and Contemporary Physics*, 146–156.

129 Friedrich Schleiermacher, *The Christian Faith* (Edinburgh: T&T Clark, 1986), 337.

Notes

[130] Robert J. Russell, "Entropy and Evil," *Zygon: Journal of Religion & Science* 19, no. 4 (1984):457.

[131] Philip Hefner, "God and Chaos: The Demiurge Versus the Ungrund," *Zygon Journal of Religion & Science* 19, no. 4 (1984):483.

[132] Harold S. Kushner, *When Bad Things Happen to Good People.*

[133] Brian Hebblethwaite, *Evil, Suffering and Religion.*

[134] E. Schillebeeckx, "Experiencing Pleasure and Anger with God's Creation," *Tijdschrift voor Theologie* 33, no. 4 (1993):325–347 (Dutch).

[135] John A. Sanford, *Evil: The Shadow Side of Reality* (New York: Crossroad, 1989), 129: "The problem of evil is unresolved in Christian theology."

[136] Harold S. Kushner, *When Bad Things Happen to Good People.*

[137] Jürgen Moltmann, *God in Creation,* 270–275. He makes some useful critical comments on the modern view of health and disease, but these do not lead to a theology of this aspect of human life.

[138] Cicely Saunders, *Care of the Dying* (London: Macmillan, 1962), 2: "...disease and all our other ills were caused in the first instance by the sin of man. These things are permitted by God because He can use them to serve His own purposes and bring about an even greater good in the end."

[139] John Rennie et al, Special Issue on Cancer, *Scientific American* 275, no. 3(September 1996): 28–119.

[140] Randolph M. Nesse and George C. Williams, "Evolution and the Origins of Disease," *Scientific American* 279, no. 5 (November 1998):58–65.

[141] Eugen Drewermann, *Exegese en Dieptepsychologie*]Exegesis and depth psychology}, 27.

[142] C.E. Lewis, C. O'Sullivan, and J. Barraclough, *Psychoimmunology of Cancer: Mind and Body in the Fight for Survival* (Oxford: Oxford University Press, 1994).

[143] Lewis Thomas, *The Youngest Science: Notes of a Medicine-Watcher* (New York: Viking Press, 1983), 204–205.

[144] Bill Moyers, *Healing and the Mind* (New York: Bantam Doubleday Dell, 1993).

[145] Sjoerd L. Bonting, *Mens, Chaos, Verzoening,* 126–127.

[146] Dale A. Matthews and David B. Larson, *The Faith Factor: Annotated Bibliography of Clinical Research on Spiritual Subjects,* 3 vols. (Rockville, MD: National Institute for Healthcare Research, 1993–95).

¹⁴⁷ *The Book of Common Prayer, Episcopal Church USA* (New York: Church Hymnal Corporation, 1979), 456: "I lay my hands upon you in the name of the Father, and of the Son, and of the Holy Spirit, beseeching our Lord Jesus Christ to sustain you with his presence, to drive away all sickness of body and spirit, and to give you that victory of life and peace which will enable you to serve him both now and evermore." A similar form has now been introduced in the Common Worship book of the Church of England (2000).

¹⁴⁸ Paul Davies, *The Last Three Minutes* (New York: Basic Books, 1994); Michael D. Lemonick, "How the Universe Will Expire," *Time*, 25 June 2001.

¹⁴⁹ Tom Gehrels, "Collisions with Comets and Asteroids," *Scientific American* 274, no. 3 (1996):34–39; John and Mary Gribbin, *Fire on Earth* (London: Simon & Schuster, 1996), 223–244. A project Spacewatch now locates and tracks all "Near Earth Objects." An impact of a 1-km asteroid (equivalent to 100,000 nuclear bombs) could destroy all or nearly all life on Earth; the chance of such an impact is estimated at once every 100,000 years. Deflecting an asteroid heading for the Earth by means of a ballistic missile with a chemical or nuclear explosive charge is being studied.

¹⁵⁰ John Macquarry, *Principles of Christian Theology* (London: SCM, 1966), 316–317.

¹⁵¹ Raymond A. Moody, *Life After Life* (New York: Bantam Books, 1975), Raymond A. Moody, *Reflections on Life After Life* (New York: Bantam Books, 1978); Maurice Rawlings, *Beyond Death's Door* (London: Sheldon Press, 1979).

¹⁵² Dorothy Sayers, "Christian Belief about Heaven and Hell," in *The Great Mystery of Life Hereafter*, ed. H.V. Hodson (London: Hodder & Stoughton, 1957), 11–18. Although she does not say this, I am sure that she also believed that this applies to all humans of any faith during their lifetime on Earth. After all, God has created us all as one human family and thus loves all of us; Christians may merely rejoice in the fact that they were privileged to know Christ already in this life and direct their life to him.

Index of Words and Names

Anglican tripos 25
Angst 30
Anselm, *Cur Deus Homo* 59
anthropic principles 42
Aquinas, Thomas 59
asteroid impact 46
Augustine 15, 22, 26, 59, 73, 74
Australopithecus 47

bará ("create") 16
Barth, Karl 17, 22, 74
big-bang theory 26, 28
bipedalism 47
Brunner, Emil 17

Calvin, John 59
Cambrian explosion 31, 46
cancer
    angiogenesis 79
    apoptosis 78
    curing or healing 81-82
    gene mutation 78
    immune system 81
    in chaos theology 80-81
    metastasis 79
    molecular mechanism 78-79
    telomere shortening 78
chance 31

chaos
    and contingency 22
    and entropy 29
    and evil 22
    initial 14, 26-27
    potential for good 23
    remaining 21, 28, 52, 76
chaos events 14, 20-26, 52, 53
chaos theology 14, 20-26, 31, 56
chaos theory 14, 32-36
chaotic thinking 23, 76
Christ, cosmic 24, 57
Clement of Alexandria 14
conservation 49, 52
contingency 22, 40
    and accidental universe 30-31
    in biological evolution 43-48
    in cosmic evolution 41-43
cosmic coincidences 41
creation 14
    continuing 21-22, 25, 49, 57, 77
    from chaos 19, 20-21
    initial 25, 56
creatio continua 22, 76
creatio ex nihilo 14, 16, 18, 19, 25, 31, 67
    biblical problem 16
    conceptual problem 15

problems with 15
scientific problem 17
theological problem 17

Dawkins, Richard 47, 71
deistic God 40, 53
demiurge 15, 20
den Heyer, C.J. 57
disease 77-82
   biblical 77-78
   cancer, molecular mechanism 78-79
   cancer, theology 79-81
   curing or healing 81-82
   sacrament of healing 82
Drees, Willem B. 35
Drewermann, Eugen 31, 80

ecological crisis 54, 76
elementary particles 28, 30
entropy 29
Enuma Elish 20
Epicurus 66
eschatology 24
evil 14, 19-20, 22-23, 66-82
   in chaos theology 76-77
   in *creatio ex nihilo* 72-75
   moral 67-69
evil matter 15, 20
evolution, biological
   contingency 43-48
   human 47-48
   punctuated 45
evolution, cosmic
   contingency 41-43

Fergusson, David 16
Fides et Ratio 67
fundamental constants 41
fundamental forces 28, 30
future
   biblical 83-84
   in chaos theology 24-25, 84
   scientific 82-83

Gnostic dualism 25
Gnostics 14, 67
God's action in the world 48-54
   immanent 50-51
   in chaos events 49
   transcendent 50
Gott, Richard 42
Gould, Stephen 31
Guth, Alan 42

Hebblethwaite, Brian 19, 75
Hefner, Philip 75
Heisenberg uncertainty principle 40
Houtepen, Anton, 74
human ambivalence 65
human evolution
   contingency 47-48
   end of 48

Incarnation 23, 55-57
inflation theory 42
information theory 29
initial mystery 26-27
intelligent design 31
Irenaeus 15, 69, 73

John Paul II 67
judgment 83

Index

kenosis 18, 24
Kuitert, H.M. 73
Kushner, Harold 66, 75, 76

Logos, creative 23
   Incarnation 55
Luther, Martin 59

Marcion 14-15, 67
miracle 53
Moltmann, Jürgen 18, 76
Monod, Jacques 31, 46
Montefiore, Hugh 51, 52
Moore, Aubrey 51
multi-world hypothesis 42

Newton, Isaac 40
Nichtige, das 17, 22, 76
Nihil 15, 17, 18
   annihilating nothingness 18, 19, 76
   nihil negativum 16
   nihil ontologicum 15
   nihil privativum 17

Origen 26, 59
original sin 69-72

panentheism 52
pantheism 52
Peacocke, Arthur 17, 34, 51
Planck time 26
Polkinghorne, John 16, 17, 34, 40, 52, 54
predestination 72
privatio boni 19, 22, 73, 76
process theology 26, 52, 54
progressiveness in evolution 31-32

reconciliation 57-61
   biblical view 57-58
   den Heyer, C.J. 57
   in chaos theology 60-61
   theological elaboration 59-60
remaining chaos 21, 28, 52, 76
Robinson, John 55
Russell, Robert 75

Sanford, John 75
Saunders, Cicely 77
Schillebeeckx, Edward 75
Schilling, Harold K. 40
scientific world view 40
separation 27
shekinah 18
singularity 26

Teilhard de Chardin, Pierre 29
theodicy 23, 77
Theophilus of Antioch 14, 15, 67
Tillich, Paul 18, 30
tohu wabohu 14
tov (Hebrew for "good") 21

universe
   accidental 30
   flat 41

virgin birth 55
Vriezen, Th.C. 26

Ward, Keith 18, 54
Westermann, Claus 16
Whitehead, Alfred N. 54
Worthing, Mark 17, 18, 74

zimsum 18

Biographical Note

Sjoerd L. Bonting was born in Amsterdam in 1924. After studying biochemistry at the University of Amsterdam (B.Sc. 1943; M.Sc. 1950 cum laude; Ph.D 1952), he moved to the United States in 1952 as National Institutes of Health (NIH) Fellow. From 1952 to 1960 he worked at the State University of Iowa, the University of Minnesota, and the University of Illinois. He was section chief at the NIH, Bethesda, from 1960 to 1965. That same year he was appointed professor and chairman, Department of Biochemistry, University of Nijmegen, the Netherlands. His main research topics were biochemistry of vision and active sodium transport. Bonting returned to the United States in 1985 as a scientific consultant for NASA at Ames Research Center, Mountain View, California, for preparation of biological research on the International Space Station. He remained in the United States until 1993.

Dr. Bonting studied theology (1957–1963) and was ordained priest at Washington Cathedral in 1964. He worked as Anglican chaplain for English-speaking persons in the Netherlands for over 20 years. After his return from the United States in 1993, he again became active in this work. In addition, he began to write extensively on science and theology. He has published 363 scientific papers and edited eight books in addition to editing *Advances in Space Biology and Medicine* (seven volumes, 1989–1999). He is the author of *Creation and Evolution: Attempt at Synthesis* (Dutch; Kok, Kampen, 1996, 2nd edition 1997); *Humanity, Chaos, Reconciliation* (Dutch; Kok, Kampen, 1998); and "Chaos Theology: A New Approach to the Science–Theology Dialogue" in *Zygon: Journal of Religion and Science* (June 1999). He is also a frequent lecturer on these topics.

Other books in the Saint Paul University Research Series: Faith and Science

Growing in the Image of God
By Carol Rausch Albright

"As persons of the early twenty-first century, we must ask: What emerging understandings are shaping our intellectual milieu and the larger culture today? When we approach questions regarding the meaning and direction of our life, we must ask how such new understandings may be shaping our images of human beings and, if we are theists, our understandings of the Originator and Sustainer of our universe. Finally, we may ask how these understandings may shed light on how best to live our lives, now, in the era we inhabit and help to shape.

"This book aims to trace two such overarching trends, and to see where they may be taking our thinking about ourselves and our responsibilities. It also asks whether, and how, such insights might influence our understanding of the Image of God."

Carol Rausch Albright, a Phi Beta Kappa scholar, is a widely published author and editor on medical and scientific subjects. She is currently co-director, Midwest region, of the Center for Theology and the Natural Sciences. Her recent publications include (with James B. Ashbrook) *The Humanizing Brain: Where Religion and Neuroscience Meet* and articles in *Zygon: Journal of Religion and Science*, of which she was executive editor from 1989 to 1998.

Crucible of Creativity: Knowing God and Nature in a Postmodern World
By Jitse M. van der Meer

"Ever since St. Augustine, an important question in Western Christianity has been how faith and reason are related. Strangely, up to a decade ago, most of what was said and written about religion and science ignored the person, in whom knowing and believing are combined.

"I take a different approach by focussing on the person as creature. This is intended as a fundamental shift away from the rationalistic framework in terms of which religion and science were seen to be interacting for the past two thousand years. I argue that they are related in the person because it is the person who stands in relation to God and nature. I assess the role of some of the faculties of humankind that are involved in the engagement of religious beliefs and the explanation of natural phenomena. Among these faculties are cognition and religious believing. I consider them as natural characteristics with which we have been created. My key point is that these faculties are related via a third faculty, that of the imagination. Together, these three faculties are 'a crucible of creativity.'"

Jitse M. van der Meer received a Ph.D in biology from the Catholic University of Nijmegen in 1978. After postdoctoral research and positions at Heidelberg and Purdui, Dr. van der Meer came to Redeemer College, Ancaster, Ontario, where he is currently professor of biology. Dr. van der meer was the founding director of the Pascal Centre for Advanced Studies in Faith and Science at Redeemer College and is its acting director.

Towards a Theology of Science
By Donald J. Lococo, CSB

"We move 'toward' a theology of science because such a discipline does not yet exist. Science assumes that theology is a subjective world view that has nothing to do with 'real' knowledge; theology is baffled by the rapid advance of natural science on the basis of presuppositions that seem to overturn its most sacred beliefs. In circles where efforts are made to bridge the divide between science and religion, theology is often peripheral to the dialogue; the emphasis is placed on religion as an empirical phenomenon that can be justified according to the criterion of rationality of natural science.

"My remarks are mostly methodological. Drawing on the theology of Hans Urs von Balthasar and the hermeneutical philosophy of Hans-Georg Gadamer, I present the concept of *logos*, the notion of a rational ground of being, as a common presupposition of science and theology. I first sketch briefly a theological interpretation of the history of Western science as a history of *logos*: the origins of the concept in Greek philosophy; its enlargement in Christian theology; its reduction to the notion of numerical unity as natural science progressively emancipated itself from theology; and finally, the forgetting of the concept in the current postmodern fragmentation of knowledge. I then examine some of the methodological issues that divide theology and natural science, in particular the necessary methodological agnosticism of empirical science, the problem revelation poses for science as an argument from authority, the need for a renewed metaphysics to mediate the agnosticism of science with the faith of theology, and the possibilities opened up for a science–theology dialogue by Vatican II."

Donald Joseph Lococo, a priest of the Basilian order, is Assistant Professor of Christianity and Culture at the University of St. Michael's College in Toronto. Dr. Lococo received a Ph.D in Zoology from the University of Toronto in 1985. He has since taught biology at the University of St. Thomas in Houston, Texas, and has published in both scientific and theological journals.

To order, contact Novalis at 1-800-387-7164 or cservice@novalis.ca

Printed and bound
in Boucherville, Quebec, Canada by
MARC VEILLEUX IMPRIMEUR INC.
in February, 2002